Application of Number
Intermediate

GNVQ resources from Stanley Thornes

GNVQ Advanced Business
GNVQ Intermediate Business

GNVQ Advanced Leisure and Tourism
GNVQ Leisure and Tourism Assignments – Book 1
GNVQ Leisure and Tourism Assignments – Book 2

GNVQ Advanced Health and Social Care
GNVQ Intermediate Health and Social Care (due Spring 1995)
GNVQ Intermediate Health and Social Care Assignments

GNVQ Built Environment Assignments – Pack 1
GNVQ Built Environment Assignments – Pack 2

GNVQ Hospitality and Catering Assignments – Pack 1 ⎫
GNVQ Hospitality and Catering Assignments – Pack 2 ⎭ (due Spring 1995)

GNVQ Core Skills in Information Technology – Level 2
GNVQ Core Skills ikn Information Technology Assignments – Pack 1
GNVQ Core Skills in Information Technology Assignments – Pack 2

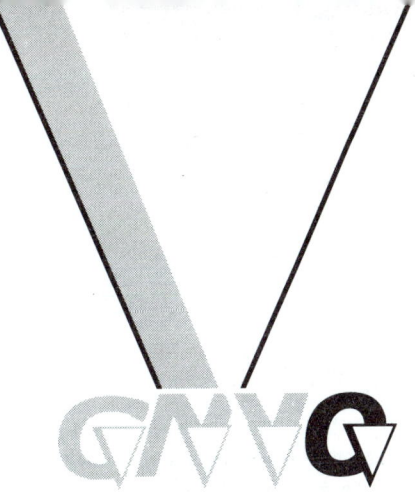

Application of Number

Intermediate

Diane Parker

GLOUCESTERSHIRE COLLEGE OF
ARTS & TECHNOLOGY

Stanley Thornes (Publishers) Ltd

Text © Diane Parker, 1994

Original line illustrations © Stanley Thornes (Publishers) Ltd 1994

The right of Diane Parker to be identified as author of this work has been asserted by her in accordance with the Copyright, Designs and Patents Act 1988.

All rights reserved. No part of this publication may be reproduced or transmitted in any form or by any means, electronic or mechanical, including photocopy, recording or any information storage and retrieval system, without permission in writing from the publisher or under licence from the Copyright Licensing Agency Limited. Further details of such licences (for reprographic reproduction) may be obtained from the Copyright Licensing Agency Limited, of 90 Tottenham Court Road, London W1P 9HE.

First published in 1994 by:
Stanley Thornes (Publishers) Ltd
Ellenborough House
Wellington Street
CHELTENHAM
GL50 1YD
UK

A catalogue record for this book is available from the British Library.

ISBN 0 7487 1791 9

Typeset by P&R Typesetters, Salisbury
Printed and bound in Great Britain at The Bath Press, Avon

Acknowledgements

The publishers are grateful to the following organisations whose material has been reproduced in this publication:

Ordnance Survey: map extract, p. 76, reproduced with the permission of the Controller of Her Majesty's Stationery Office © Crown Copyright
Central Statistical Office: extracts, pp. 114, 133, 137, 141, 142, 144, 146, 147, 148, 149, 167 from *Social Trends* © Crown Copyright.
British Rail: timetable extracts, pp. 114, 120
Halifax Building Society: loan repayment table, p. 115
Gloucestershire Echo: currency exchange rate information, p. 115
The Independent: currency exchange rate information, p. 115; accident data, p. 150
Forte Leisure Breaks: hotel data, p. 116
O.E.C.D.: house building data, p. 124; education enrolment data, p. 149
Heart of England Tourist Board: occupancy data, p. 136
Department of Employment, data on employment, p. 137
Dean Forest Studios: photographs pp. 169, 195
London Regional Transport: underground map, p. 229
Gloucestershire Airport: timetable, p. 237

Every effort has been made to contact copyright holders but the publishers apologise if any have been overlooked.

Contents

Introduction   vi

Module 1   Number

1   **Estimating and Calculating** (Core Skill 2.1)   **1**
2   **Directed Numbers** (Core Skill 2.2)   **12**
3   **Fractions** (Core Skill 2.2)   **32**
4   **Decimals** (Core Skill 2.2)   **53**
5   **Ratios, Scales and Percentages** (Core Skill 2.2)   **72**
6   **Conversions** (Core Skill 2.1)   **88**

Module 2   Handling Information

7   **Surveys and Questionnaires** (Core Skill 2.1)   **102**
8   **Tabulating your Information** (Core Skill 2.1)   **113**
9   **Illustrating Information** (Core Skill 2.3)   **122**
10   **Measuring Information** (Core Skill 2.3)   **154**
11   **Probability** (Core Skill 2.3)   **168**

Module 3   Shape and Space

12   **Perimeters and Areas** (Core Skills 2.2, 2.3)   **179**
13   **Volumes** (Core Skills 2.2, 2.3)   **209**
14   **Networks – what are they?** (Core Skill 2.2)   **226**

Module 4   Language of Mathematics

15   **Formulae and Equations** (Core Skill 2.2)   **239**

Solutions to Further Questions   253

Index   266

Introduction

When GNVQ students join their vocational courses they find that maths, in the guise of the core skills Application of Number, is a requirement of the course. Having experienced the National Curriculum and GCSE the student does not want a repetition of traditional teaching methods. Each student will have different strengths and weaknesses and therefore a new approach is needed to capture their imagination.

This book aims to provide a new approach to the delivery of Application of Number. It is written for the student, not for the lecturer. It is designed to be student-friendly, trying to identify the common problems experienced with number skills. Once the problem has been identified the student is guided to a solution which is based on understanding.

The book is written as four modules:
- Number
- Handling Information
- Shape and Space
- Language of Mathematics

Each module is made up of a number of chapters which contain exercises of general and vocational interest. Full solutions are provided to help the student to identify problems and to demonstrate correct layout. Each chapter concludes with a summary of what has been learnt followed by further vocationally-oriented questions.

All students use calculators but very few know how to make the best use of this powerful tool. This book encourages the use of calculators and shows the student which keys to use and demonstrates how they should be used. It also helps students to understand when particular keys should be used. This is an important skill which students will find very useful in their vocational studies.

I should like to thank my children, Sarah and David, for their encouragement and help. Without them, this book would not have been written.

1 Estimating and Calculating

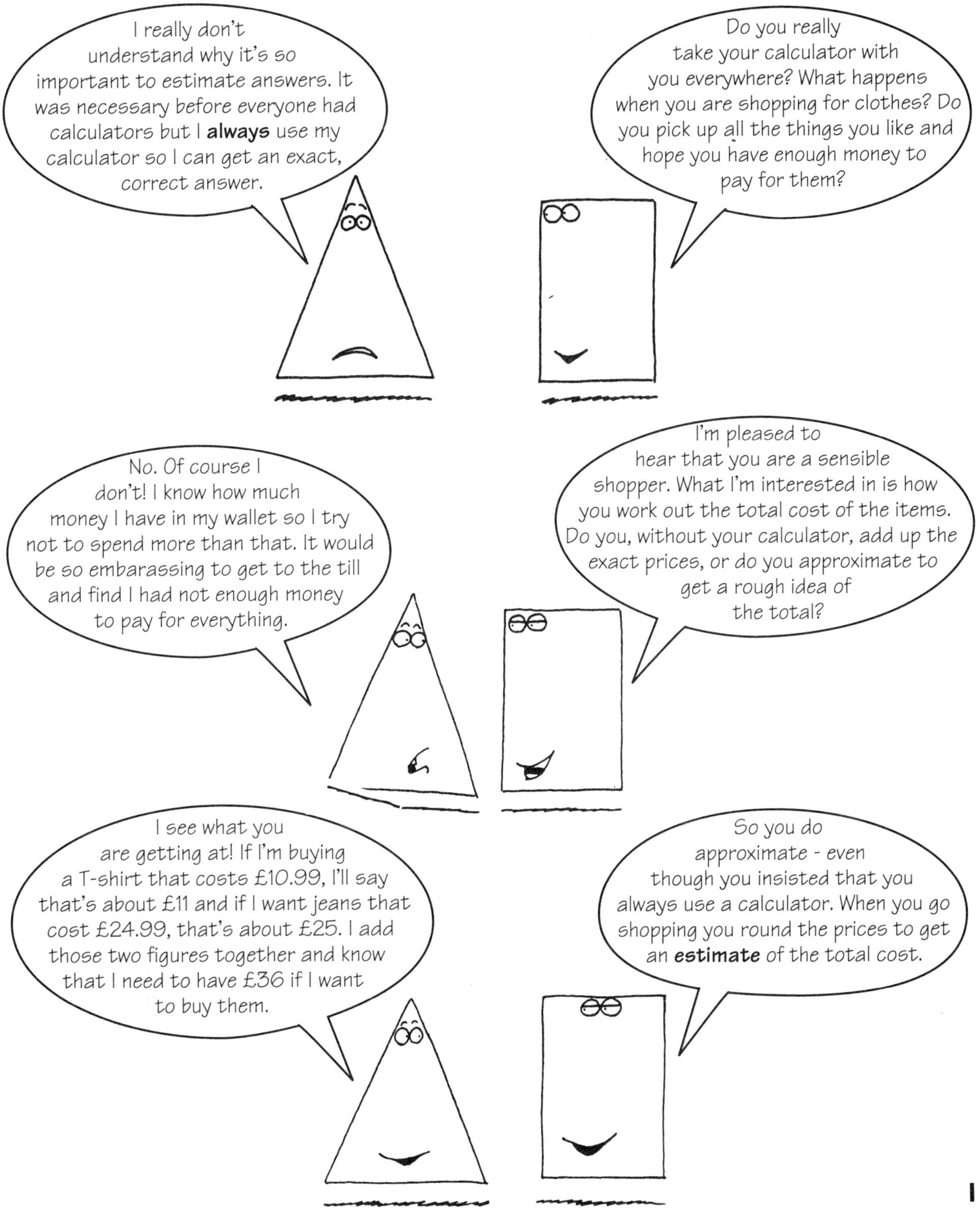

I really don't understand why it's so important to estimate answers. It was necessary before everyone had calculators but I **always** use my calculator so I can get an exact, correct answer.

Do you really take your calculator with you everywhere? What happens when you are shopping for clothes? Do you pick up all the things you like and hope you have enough money to pay for them?

No. Of course I don't! I know how much money I have in my wallet so I try not to spend more than that. It would be so embarassing to get to the till and find I had not enough money to pay for everything.

I'm pleased to hear that you are a sensible shopper. What I'm interested in is how you work out the total cost of the items. Do you, without your calculator, add up the exact prices, or do you approximate to get a rough idea of the total?

I see what you are getting at! If I'm buying a T-shirt that costs £10.99, I'll say that's about £11 and if I want jeans that cost £24.99, that's about £25. I add those two figures together and know that I need to have £36 if I want to buy them.

So you do approximate - even though you insisted that you always use a calculator. When you go shopping you round the prices to get an **estimate** of the total cost.

1

Rounding

Let's look at an easy rule to help you to round numbers.

These jeans cost £49.99, which is about £50. £49.99 is much nearer to £50 than to £40, so we round *up* the price to £50.

This pair of jeans costs £32.99. That is much nearer to £30 than to £40. This time we round *down* to £30.

Both prices have been rounded to the nearest £10.
When we round to the nearest 10, the number always ends in zero (0).
The rule you use is very easy to remember.

```
If the last digit is
   1 2 3 4        5 6 7 8 9
   round down     round up
```

```
The numbers
1, 2, 3, 4, 5, 6, 7, 8, 9, 0
are called digits.
```

 Try some for yourself

1. Round these numbers to the nearest 10.

7, 34, 105, 98, 13.49

Rounding to the nearest 100

When we rounded to the nearest 10 we looked at the last digit and then used our rule to round up or down. When we round to the nearest 100, we look at the last *two* digits and then round up or down.

235 is nearer to 2100 than 300.   35 is less than 50 so round down to 200.
268 is nearer to 300 than 200.   68 is more than 50 so round up to 300.

 Try some for yourself

2. Round these numbers to the nearest 100.

499.50, 199, 129.90, 302.6

Investment manager stole £645,000 from his clients

Have you noticed how newspaper headlines always contain nice round numbers? This is because from the reader's point of view £645 000 gives a good idea of the size of the theft. The manager could have stolen anything from £644 500 to £645 449 but £645 000 gives an immediate idea of the scale of the theft.

Rounding is often used with large numbers to give an idea of the order of magnitude (the rough size) of the numbers involved.

Here's a quick guide for rounding numbers.

Nearest 10	1, 2, 3, 4	5, 6, 7, 8, 9	1 zero – look at the last digit
Nearest 100	1–49	50–99	2 zeros – look at the last 2 digits
Nearest 1000	1–499	500–999	3 zeros – look at the last 3 digits
Nearest 1 000 000	1–499 999	500 000–999 999	6 zeros – look at the last 6 digits

(Less than half-way? Round down.)   (Half-way or more? Round up.)

There are plenty of practical examples to look at.

Travel

The distance from London to Aberystwyth is 212 miles. Round the distance to the easiest number to handle (one digit and lots of noughts).

We need to round to the nearest 100.
212 rounds *down* to 200.

So the distance from London to Aberystwyth is approximately ($\simeq$) 200 miles.

Shopping

We need to estimate the cost of the grocery bill so we must round the cost of each item.

£0.79 is approximately £1.00
£1.49 is approximately £1.50
£0.45 is approximately £0.50
£2.99 is approximately £3.00
£0.69 is approximately £0.70
£1.99 is approximately £2.00

Total £8.70

	£
CIABATTA BREAD	0.79
SPAGHETTI	1.49
SEAFOOD STICKS	0.45
T.PRAWNS 400G	2.99
DK MUSCOVADO	0.69
FLOWERS	1.99

 Try some for yourself

3. (a) John has interviews at colleges in Southampton, Bristol and Birmingham on consecutive days. He wants to gain a rough idea of the total mileage for the round trip from his home in London and back again.

London–Southampton	77 miles
Southampton–Bristol	62 miles
Bristol–Birmingham	121 miles
Birmingham–London	110 miles

Find the approximate mileage *and* the exact mileage.

(b) The number of British servicemen killed in World War I was 908 371. Round this number off to the easiest number to work with – one digit and lots of noughts.

(c) Jane has offered to go to the shop to buy cans of drink for herself and five friends. The cans cost 37 pence each. Make a rough estimate of the cost of the six cans.

(d) Make a rough estimate of the total cost of this supermarket bill.

	£
2LTR S+S ORANGE	1.99
LTR F/S ORNG	1.99
700ML APPLESPRK	0.99
RTB TOM CIABAT	1.49
GRAPE WHITE /LB	1.94
SWISS B/C YOG	0.35

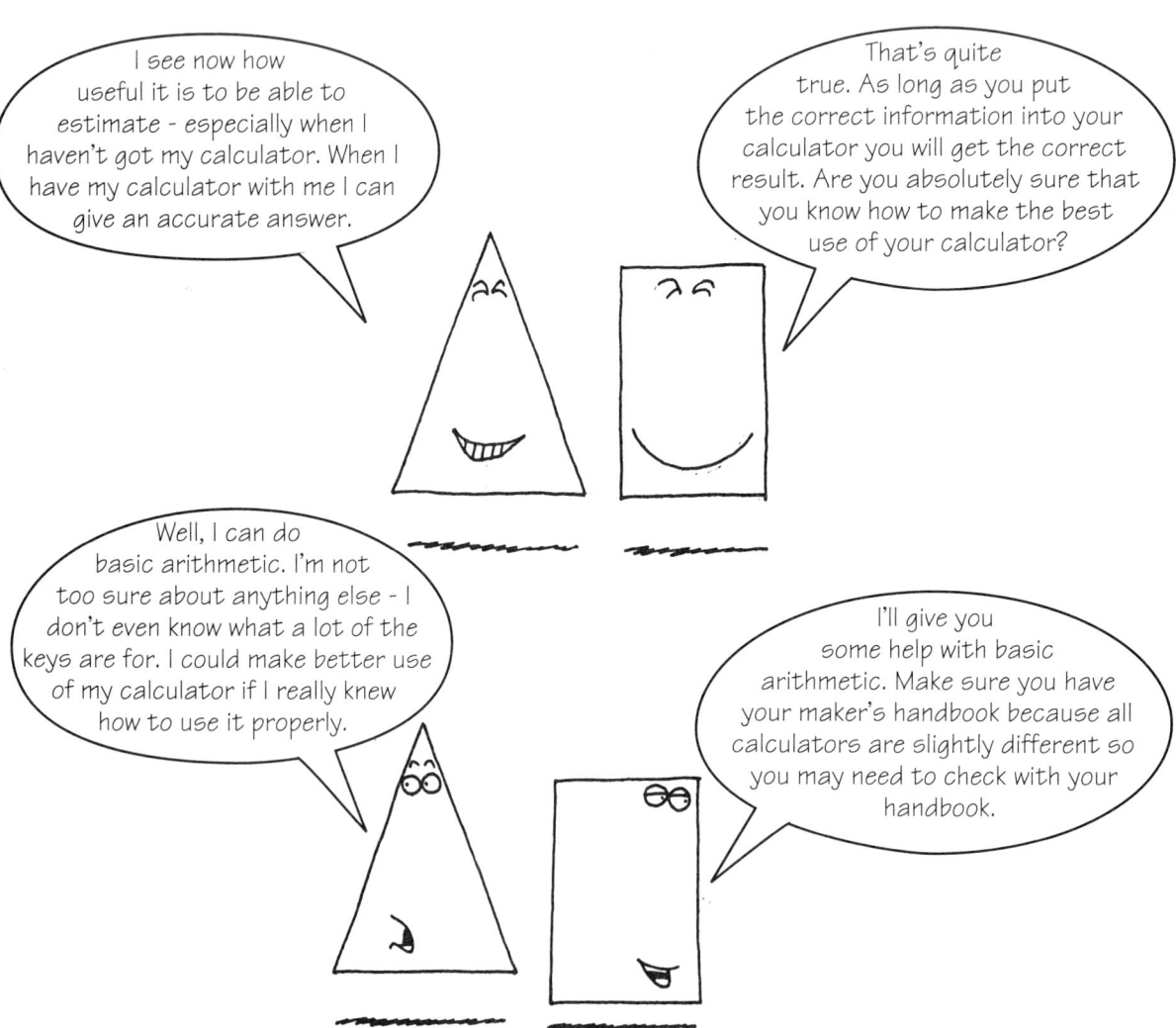

We'll begin by looking at the basic keys found on all calculators.

Using your calculator

In many areas of work a calculator is a necessary evil. It makes calculations swift and simple and it is quite easy to use once you become familiar with its operation.

There are many different makes of calculator and many different ways of using one. The important thing is for you to get used to your calculator. The first step is to make sure that you have the maker's handbook for your calculator as you will need to refer to it quite frequently at first. With the handbook beside you, work through the following activities and you will soon acquire confidence.

First, look for the following features:

$\boxed{\text{ON}}$ $\boxed{\text{OFF}}$ The on/off switch, possibly the most important feature particularly if the calculator is battery operated.

$\boxed{\text{AC}}$ The 'clear' key. This 'clears' the display. It can be labelled $\boxed{\text{AC}}$ or $\boxed{\text{C}}$. Check with your maker's handbook.

$\boxed{+}$ $\boxed{-}$ $\boxed{\times}$ $\boxed{\div}$ These are the 'operations' keys.

$\boxed{=}$ The equals key.

$\boxed{0}$ $\boxed{1}$ $\boxed{2}$ $\boxed{3}$ $\boxed{4}$ $\boxed{5}$ $\boxed{6}$ $\boxed{7}$ $\boxed{8}$ $\boxed{9}$ The digits.

Your calculator will have other keys, such as function keys or memory keys. For the moment we shall concentrate on mastering the keys listed above.

Let's try entering numbers into the display. Remember to clear the display each time.

Try entering 21, 32, 154, 09, . . .

Key sequence

When you perform a specific calculation on your calculator you will need to follow a **key sequence**. A key sequence is just a list of keys to press in sequence.

For example, if you want to calculate 2 + 7 you follow the key sequence

$\boxed{2}$ $\boxed{+}$ $\boxed{7}$ $\boxed{=}$

Start from the left-hand side then work across, pressing each key in turn.

 Try some for yourself

4. Write down the key sequence for each of the following calculations. Do the calculations on your calculator and check the answer in your head or on paper. This will build up your confidence in the way your calculator works and will enable you to proceed to more complicated calculations.

(a) 6 + 4 = (d) 9 ÷ 3 =
(b) 7 − 3 = (e) 5 × 2 =
(c) 4 × 6 = (f) 8 ÷ 2 =

The same process can be used for harder calculations.

Example

Calculate 45 737 − 33 897

Solution

The key sequence is

[4][5][7][3][7][−][3][3][8][9][7][=]

Display

Remember that a calculator is useful only if you can rely on the answer. When you use one you should stop and think whether the answers you get are reasonable, because it is quite easy to press the wrong key or make some other mistake. One way of checking your answer is to round off the numbers and make a rough estimate.

In the example above the answer can be checked as follows:

45 737   rounds up to   50 000
33 897   rounds down to   30 000

50 000 − 30 000 can be done in your head – it gives 20 000. This is not very close to the answer but it does indicate that the answer should be a five-digit number less than 20 000.

 Try some for yourself

5. (a) For each of the following, make a rough estimate and then find the exact answer using your calculator. Write down the key sequence first.

    (i) 324 + 89   (iv) 10 896 074 − 8 998 894
    (ii) 67 534 − 18 693   (v) 6354 − 2547
    (iii) 819 116 + 706 347   (vi) 21 000 001 − 19 894 020

(b) Some of the following answers are wrong as the result of mistakes in keying the calculations into the calculator. Find which are the wrong ones, without using your calculator.

    (i) 521 − 468 = 989   (iii) 1 854 361 − 974 807 = 879 554
    (ii) 2176 + 901 = 2177   (iv) 635 827 − 364 107 = 1720

Quite often the estimate is not a particularly good approximation to the answer. This happens most frequently when the numbers are of different sizes, when this is the case, you should round the numbers to the same level of accuracy.

Example

Evaluate   74 132 − 637

Solution

Estimate first − round both numbers to the same level of accuracy, the nearest 1000.

74 132   rounds down to   74 000
673   rounds up to   1000

74 000 − 1000 = 73 000

Then use your calculator

74 132 − 673 = 73 459

The estimate indicates that the answer would be a five-digit number close to 73 000. The estimate is said to indicate the **order of magnitude** of the answer.

> **The order of magnitude**
> - indicates the number of digits in the answer
> - gives a rough idea of the size of the answer.

 Try some for yourself

6. For each of the following, first work out the order of magnitude of the answer then calculate it exactly.

(a) 652 − 13   (c) 15 712 + 456
(b) 1934 − 156   (d) 19 714 + 1554

Multiplication

Example

Evaluate   2376 × 4235

Solution

Estimate first

2376   rounds down to 2000   = 2 × 1000
4235   rounds down to 4000   = 4 × 1000

2000 × 4000 = 2 × 4 × 1000 × 1000
                        3 zeros 3 zeros

= 8 × 1 000 000
        6 zeros

= 8 000 000

Calculate second

2376 × 4235 = 10 062 360

The estimate is not very close, but it does indicate the order of magnitude.

Try some for yourself

7. For each of the following, first make a rough estimate then find the exact answer using your calculator.

   (a) 325 × 184
   (b) 6213 × 2617
   (c) 15 038 × 577
   (d) 724 × 136
   (e) 3079 × 171
   (f) 111 × 903

Using the calculator to solve problems

By now, you should be gaining in confidence in using your calculator for straightforward calculations. When we look at everyday situations we need to decide

- when to use the calculator
- whether the answer makes sense in the context of the problem.

Example

David's car does 38 miles to the gallon. Each week he buys 12 gallons of petrol. How far does he travel each week?

Solution

Estimate

38   rounds up to   40
12   rounds down to   10
40 × 10 = 400

Calculate

38 × 12 = 456

 Try some for yourself

8. (a) Mortgage repayments are £172 per month and Council Tax is £398 per annum. What is the yearly expenditure for this household?

   (b) What is the area of a football pitch measuring 360 ft by 240 ft?

   (c) A local firm employs 294 workers, each earning £6796 a year. What is the annual salary bill for the firm?

What have you learnt about estimation and calculation?

> I know that estimation is easy and I understand why it is important. Now I feel happier and more confident about using my calculator.

☑ I can round numbers when I need to estimate.

☑ I can use my calculator to add, subtract, multiply and divide.

☑ I can estimate the order of magnitude of a calculation to help me check the answer I get on my calculator.

☑ I can estimate and calculate when dealing with practical problems.

Further Questions

1. John and Julia decide to have a Chinese take-away. John wants to make sure that he takes enough money when he goes to collect the food. He estimates the total cost by rounding all the figures to the nearest £1. This is his order.

1 chicken and noodle soup	£0.95
1 chicken and sweetcorn soup	£1.05
1 portion egg fried rice	£1.15
1 portion Chow Mein	£1.65
Chicken with cashew nuts	£3.25
Shrimp with mushrooms	£2.85
Prawn crackers	£0.90

2. A businessman has to make a long trip to visit a number of companies. His route takes him from London to Leeds, Manchester and Birmingham before he returns to London. He needs to estimate his mileage to the nearest 10 miles.

London–Leeds	253 miles
Leeds–Manchester	72 miles
Manchester–Birmingham	126 miles
Birmingham–London	110 miles

   Estimate his total mileage by rounding all figures to the nearest 10 miles.

3. The Smith family are planning a holiday for November. They are going to stay at an hotel near Legoland. The party is made up of Mr and Mrs Smith and their son James.

   The prices given in the holiday brochure for 7 nights accommodation with flights included, are:

1 adult	£498.00
1 child	£298.80

   Estimate the cost of the holiday for two adults and one child.

4. Christopher has asked his parents for a hand-held games system and two games. His parents want an estimate of the cost of the three items.

Games system	£39.99
First game	£24.99
Second game	£29.99

5. Anisa is going on holiday in the summer. She is wondering how much she needs to allow for buying clothes for her holiday. She knows she will need these clothes and accessories:

Swimsuit	£19.99
Toe-loop sandals	£14.99
Straw bag	£12.99
Cotton V-neck top	£29.99

   Estimate the amount she needs to save, to the nearest £1.

6. The O'Connor family are going to Dublin for a weekend break. They will fly from London to Dublin. The family includes Mr and Mrs O'Connor and their sons, Shaun and Gerard. The children are under 12 years so qualify for price reductions. The prices are:

1 adult	£186
1 child	£136

   Estimate to the nearest £10, the cost of the family's weekend break.

7. Ahkter wants to try the new Indian take-away which has opened in his neighbourhood. He orders the items he wants by telephone but needs to make sure he takes enough money when he goes to collect his order. Estimate the total cost by rounding to the nearest £1.

3 plain papadums at 30p each	
1 portion Tandoori Chicken	£3.95
1 Chicken Dhansak	£3.85
Pilau rice	£1.00
1 nan bread	£0.95
1 chapati	£0.70

8. Yang has to travel from Gloucester to Cambridge, then on to Oxford before returning home. She wishes to estimate the distance she must travel, to the nearest 10 miles.

Gloucester–Cambridge	117 miles
Cambridge–Oxford	69 miles
Oxford–Gloucester	49 miles

Solutions

1. 7 is nearer to 10 than 1   so round *up* to 10.
   34 is nearer to 30 than 40   so round *down* to 30.
   105 is nearer to 110 than 100   so round *up* to 110.
   98 is nearer to 100 than 90   so round *up* to 100.
   13.49 is nearer to 10 than 20   so round *down* to 10.

2. 499.50 is nearer to 500 than 400   so round *up* to 500.
   199 is nearer to 200 than 100   so round *up* to 200.
   129.90 is nearer to 100 than 200   so round down to 100.
   302.6 is nearer to 300 than 400   so round *down* to 300.

3. (a) London–Southampton $\simeq$ 100 miles
   Southampton–Bristol $\simeq$ 100 miles
   Bristol–Birmingham $\simeq$ 100 miles
   Birmingham–London $\simeq$ 100 miles
   Total $\simeq$ 400 miles
   Exact 370 miles

   (b) 900 000

   (c) 37p $\simeq$ 40p   40 × 6p = 240p = £2.40

   (d)
£
2.00
2.00
1.00
1.50
2.00
0.50
9.00

4. (a) $\boxed{6} \boxed{+} \boxed{4} \boxed{=} 10.$
   (b) $\boxed{7} \boxed{-} \boxed{3} \boxed{=} 4.$
   (c) $\boxed{4} \boxed{\times} \boxed{6} \boxed{=} 24.$
   (d) $\boxed{9} \boxed{\div} \boxed{3} \boxed{=} 3.$
   (e) $\boxed{5} \boxed{\times} \boxed{2} \boxed{=} 10.$
   (f) $\boxed{8} \boxed{\div} \boxed{2} \boxed{=} 4.$

5. (a) (i) Estimate 300 + 100 = 400
   Calculate 324 + 89 = 413

   (ii) Estimate 68 000 − 19 000 = 49 000
   Calculate 67 534 − 18 693 = 48 841

   (iii) Estimate 800 000 + 700 000 = 1 500 000
   Calculate 819 116 + 706 347 = 1 525 463

   (iv) Estimate 11 000 000 − 9 000 000 = 2 000 000
   Calculate 10 896 074 − 8 998 894 = 1 897 180

   (v) Estimate 6000 − 3000 = 3000
   Calculate 6354 − 2547 = 3807

   (vi) Estimate 21 000 000 − 20 000 000 = 1 000 000
   Calculate 21 000 001 − 19 894 020 = 1 105 981

   (b) (i) Incorrect   A $\boxed{+}$ was keyed in instead of $\boxed{-}$
   521 + 468 = 989

   (ii) Incorrect 2176 + 001 = 2177
   A miskeying of 0 for 9

   (iii) Correct

   (iv) Incorrect   The number 36 was transposed to 63
   635 827 − 634 107 = 1720

6. (a) Estimate 650 − 10 = 640
   Calculate 652 − 13 = 639

   (b) Estimate 1900 − 200 = 1700
   Calculate 1934 − 156 = 1778

   (c) Estimate 15 700 + 500 = 16 200
   Calculate 15 712 + 456 = 16 168

   (d) Estimate 20 000 + 2000 = 22 000
   Calculate 19 714 + 1554 = 20 258

7. (a) Estimate 300 × 200 = 60 000
   Calculate 325 × 184 = 59 800

   (b) Estimate 6000 × 3000 = 18 000 000
   Calculate 6213 × 2617 = 16 259 421

   (c) Estimate 15 000 × 600 = 9 000 000
   Calculate 15 038 × 577 = 8 676 926

   (d) Estimate 700 × 100 = 70 000
   Calculate 724 × 136 = 98 464

   (e) Estimate 3000 × 200 = 600 000
   Calculate 3079 × 171 = 526 509

   (f) Estimate 100 × 900 = 90 000
   Calculate 111 × 903 = 100 233

8. (a) Mortgages £172 per month   round to £200
   Mortgages £172 × 12 per year
   £200 × 12 = £2400
   Poll tax £398   round to £400
   Total = £2800

   Calculate £172 × 12   £2044
   add Poll tax   £ 398
   Total   £2462

   (b) Length = 360 ft   Width = 240 ft
   Area: Estimate   400 × 200 = 80 000 ft^2
   Calculate   360 × 240 = 86 400 ft^2

   (c) One worker earns £6796   Estimate £7000
   294 workers earn   Estimate
   £6796 × 294   £7000 × 300
   Actual total = £1 998 024   Estimated total
   = £2 100 000

 Directed Numbers

My lecturer has been talking about things called directed numbers. I didn't know numbers needed to be given directions. Even worse are these negative numbers. I can't understand what I'm supposed to do with them.

Numbers do have direction – that's why we talk about directed numbers. There are two directions in which numbers can move. They can move in a positive direction – getting larger – or in a negative direction – getting smaller.

Understanding directed numbers

Numbers like 0, 1, 2, 3, . . . can be shown on a diagram if we draw a line and mark off the numbers. This is called a **number line**.

When we add one number to another we move along the number line.

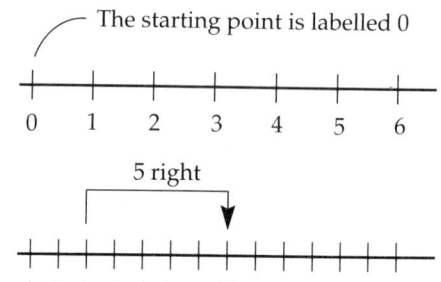

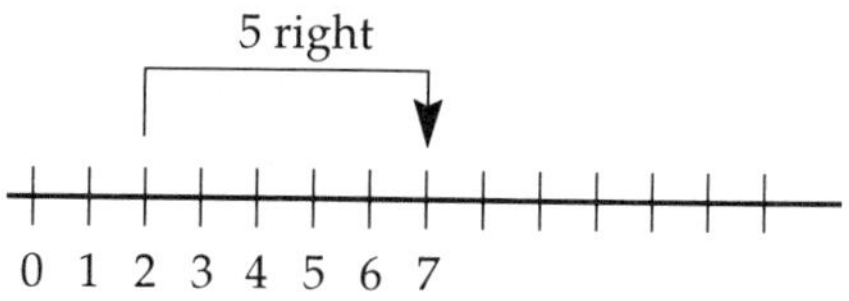

Let's look at what happens when we add 5 to 2.

Start at 2 and move 5 steps to the right. (The right is the **positive** direction).

$$2 + 5 = 7$$

That's just a fancy way of adding two numbers. What do you do with this number line when you want to take one number from another?

*To take one number from another just move to the left along the number line. The left is the **negative** direction.*

Let's take 3 from 7.

Moving 3 steps to the left, in the negative direction, brings us to 4.

$7 - 3 = 4.$

Try taking 7 from 3.

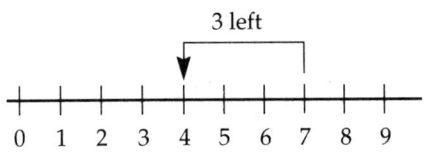

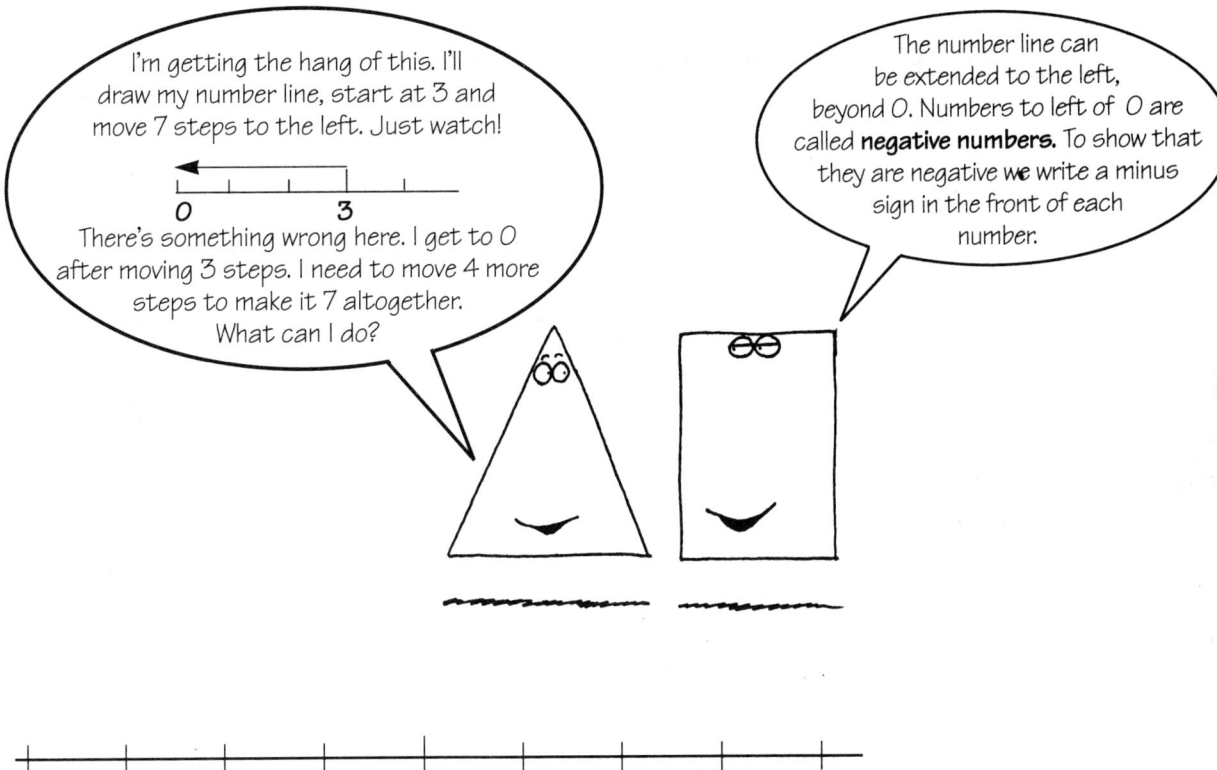

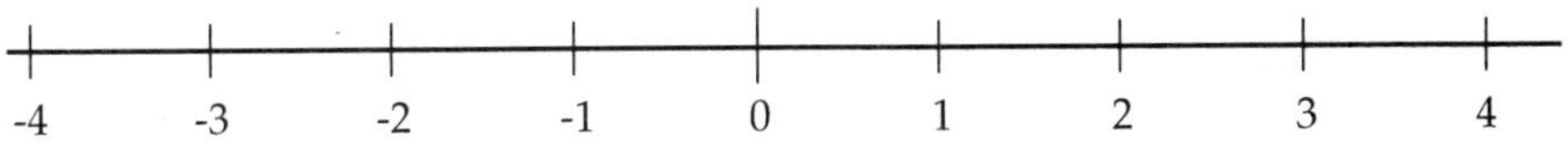

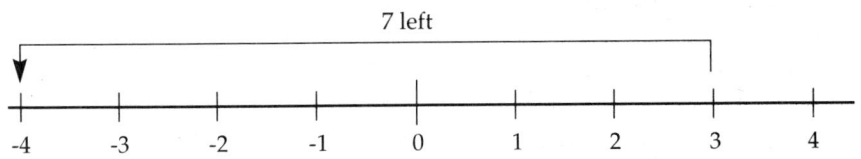

13

Try some for yourself

1. Draw a number line to help you.

(a) 4 − 4   (c) 2 − 9

(b) 3 − 8   (d) − 2 + 6

Working with directed numbers

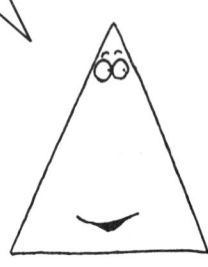
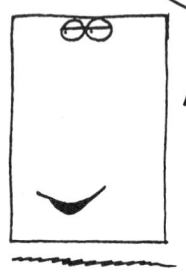

I think I understand how to use directed numbers. Tell me what practical use they have. Where will I find a practical use for negative numbers?

We often use negative numbers in everyday life, perhaps not being aware of it. The first person to use negative numbers was a Hindu mathematician, called Brahmagupta. They were used to represent debts. I'll give some examples of their use.

Bank statements

A. Mathematician			
Debit		**Balance**	
		279.13	
143.10		136.03	
50.00		86.03	
104.05		18.02	DR

DR next to the figure £18.02 means that A. Mathematician was overdrawn by £18.02. He owed the bank £18.02. In fact, as far as the bank was concerned he owned £(− 18.02) or − £18.02.

So, when you are 'in the red', it means that you have a negative bank balance.

Temperature

In winter months the temperature often drops below freezing point. 0°C is freezing point. The temperature scale is just like the number line.

(− 3)°C is 3° below freezing point.

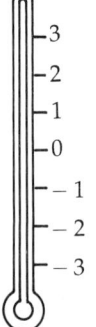

°C means degrees centigrade or Celsius

Contour maps

Contours on a map show the height of land above sea level. Modern maps show the height in metres at 50 metre intervals.

When the land is actually *below* sea level, as in some parts of Holland, the corresponding contours indicate negative heights.

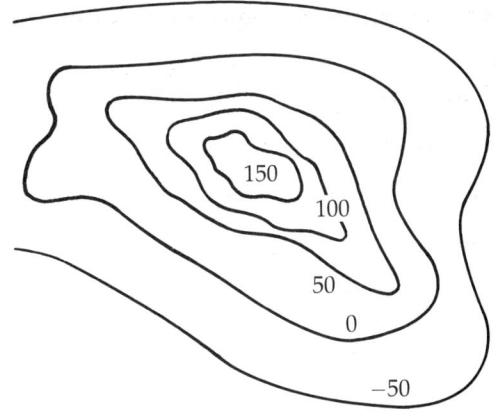

I can see how to move along the number line. I go to the right with a positive number and to the left with a negative one. What happens when I want to use directed numbers to solve problems?

You can work with negative numbers just as you can work with positive numbers. You may find that the results are not what you expect. People often think negative numbers are a little 'magical'. Strange things happen when they are subtracted, multiplied, added and divided.

Remember

- Positive numbers lie to the right of 0 on the number line.
- Negative numbers lie to the left of 0 on the number line.
- The number 0 is very special. It is neither negative nor positive.
- Numbers can be positive, negative or zero.

A calculator helps

I don't like this talk of magic. I need my maths to be straightforward. I want to be given rules that I understand. Most of all I want to be able to use my calculator with these 'magical' numbers.

Using your calculator is a good idea. You will be able to discover for yourself the rules governing negative numbers. That will help you to understand these magical numbers.

How to enter a negative number into your calculator

Your calculator should have a key marked like this $\boxed{+/-}$. This is called a **change of sign** key. This key changes the sign.

Try entering $\boxed{3}\boxed{+/-}$

The display should give -3.

Change the sign again $\boxed{+/-}$

Now the display gives 3.

 Try some for yourself

2. Enter each of the following numbers into your calculator.

(a) -4 (b) -6 (c) -1782 (d) -2941

We started by using the number line. Then we used a calculator. How does this relate to the number line?

I'm glad you have not forgotten the number line. We are now going to use the number line and the rule you found.

We start at -3 and move 1 place to the left, in the negative direction.

$(-3) - 1 = -4$

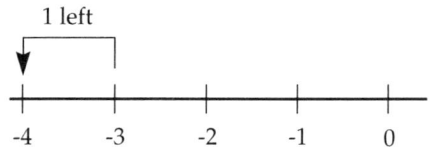

The rule for adding negative numbers tells us that

$$(-2) + (-4) = (-2) - 4$$

Now we can use the number line.

Start at (-2) on the number line and move 4 left.

$(-2) - 4 = -6$
so $(-2) + (-4) = -6$

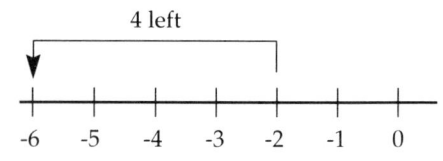

 Try some for yourself

3. (a) $(-3) - 7$ (c) $2 + (-6)$ (e) $(-3) + (-8)$
 (b) $(-2) - 9$ (d) $4 + (-7)$ (f) $(-1) + (-2)$

Adding negative numbers

That's quite easy. I did not know what that key was for — I've learned something! Now, how about something more difficult? Can I do some sums now?

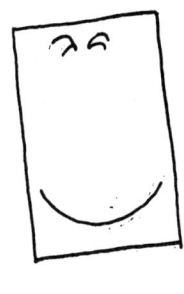

I think you are ready to tackle a mathematical investigation to find out what happens when you **add** a negative number.

 Try some for yourself

4. Use your calculator to work out each of the following. For each question compare your answer to part (i) with your answer to part (ii).

(a) (i) 3 + (− 4)   Key sequence 3 + 4 +/− =
   (ii) 3 − 4   Key sequence 3 − 4 =
(b) (i) 3 + (− 10)   (ii) 3 − 10
(c) (i) 2 + (− 7)   (ii) 2 − 7
(d) (i) 5 + (− 8)   (ii) 5 − 8

For each question the answer to part (i) was the same as the answer to part (ii).
3 + (-4) = -1   3 - 4 = -1

Now I want you to see what happens if the first number is negative. Will you find the same pattern?

 Try some for yourself

5. (a) (i) (− 2) + (− 7)   (ii) (− 2) − 7
   (b) (i) (− 3) + (− 11)   (ii) (− 3) − 11

17

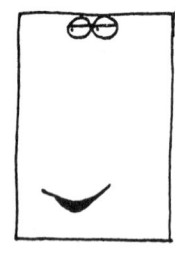

The same thing happened!
-2 + (-7) = -9 and -2 - 7 = -9
I'm sure that must tell me something but I'm not sure how to explain it.

You have just discovered the rule about adding negative numbers.

Adding a **negative** number is the same as **subtracting** a **positive** number.

Subtracting negative numbers

Now I want you to see what happens when you subtract negative numbers. I want you to try the following subtractions using your calculator.

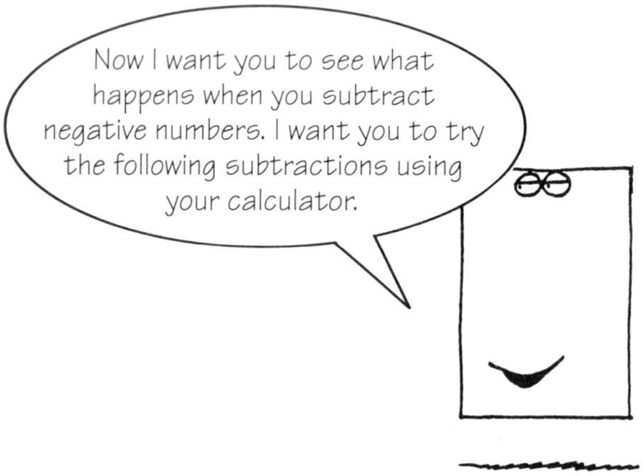

Try some for yourself

6. (a) $2 - (-3)$ (d) $8 - (-10)$
    (b) $4 - (-7)$ (e) $1 - (-1)$
    (c) $6 - (-2)$

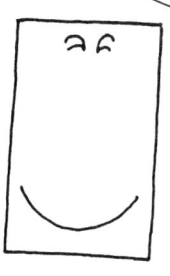

This is fun. When I subtract a negative number I get the same answer that I would have if I added a positive number. That's magic!

You have discovered the rule for subtracting negative numbers. This is really good work.

Subtracting a **negative** number is the same as **adding a positive** number.

Practical applications

 Try some for yourself

7. Sarah is £36 overdrawn. How much must she add to her bank account to be out of debt?

8. John was rock climbing in Israel. He started at 53 metres below sea-level and ended up 20 metres above sea-level. How far had he climbed?

9. The temperature was $-3°C$ on Tuesday and dropped to $-10°C$ on Wednesday. But how much had it dropped?

10. Andrew had £13 in his bank account. He cashed a cheque for £25, then another for £20. How much did he have left after these transactions?

Multiplying negative numbers

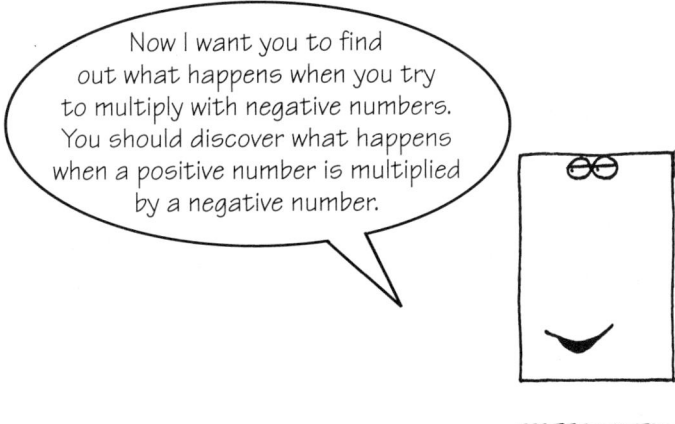

19

Try some for yourself

11. Work out the following. Use your calculator for part (i); you should be able to use your head for part (ii).

Is the answer positive or negative?

(a) (i) 2 × (−3)   Key sequence [2] [×] [3] [+/−] [=]
    (ii) 2 × 3

(b) (i) 4 × (−7)   (ii) 4 × 7

(c) (i) 6 × (−10)   (ii) 6 × 10

(d) (i) 1 × (−1)   (ii) 1 × 1

When I multiply a positive number by a negative number the answer is negative. When I multiply a positive number by a positive number the answer is positive.
+ × − = −   + × + = +

That's right – you are beginning to discover the rule. Now I want you to find out what happens when you multiply a negative number by a positive number.

 Try some for yourself

12. Try these examples. The numbers are the same as the previous questions. The signs are different.

(a) (−2) × 3   (c) (−6) × 10
(b) (−4) × 7   (d) (−1) × 1

The signs may be different but I get the same answers. It doesn't matter if the negative number comes first or last, the answer is always negative.
− × + = −

20

This suggests that

(a positive number) × (a negative number) = (a negative number)

((+) × (−) = (−))

and

(a negative number) × (a positive number) = (a negative number)

((−) × (+) = (−))

You are doing very well. There is just one more combination to investigate. You need to find out what happens when you multiply a negative number by another negative number.

 Try some for yourself

13. Try these examples. Again, the numbers are the same, but the signs are different. Is the answer positive or negative?

(a) $(-2) \times (-3)$ (c) $(-6) \times (-10)$
(b) $(-4) \times (-7)$ (d) $(-1) \times (-1)$

All the answers were positive. Is that right, or did I make a mistake? If it is right can you explain why it happens? − × − = +

You are quite correct, no mistakes. At this level it is difficult to explain **why** it works in this way. This is why negative numbers seem 'magical'. The result is rather mysterious.

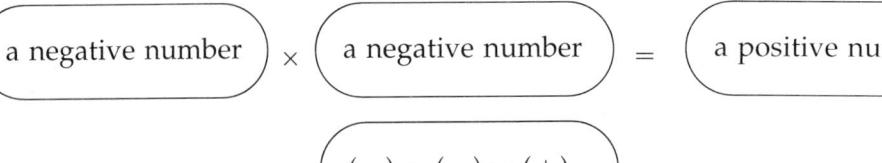

a negative number × a negative number = a positive number

(−) × (−) = (+)

If the signs are the **same** − the answer should be **positive**.
If the signs are **different** − the answer should be **negative**.

Dividing negative numbers

My eyes tell me that when I multiply two negative numbers I get a positive answer. I believe what I see but I don't understand. What happens with division? Division and multiplication are opposites − do the same rules apply?

I'm really sorry that I cannot give a simple explanation for the rules about multiplying negative numbers. Try these examples and deduce the rule for division. Believe what you see!

Try some for yourself
14. (a) $20 \div (-4)$
(b) $-36 \div (-12)$
(c) $-42 \div 14$
(d) $81 \div (-9)$
(e) $-15 \div 3$
(f) $-27 \div (-3)$
(g) $-39 \div (-13)$
(h) $-504 \div 56$
(i) $12 \div 4$

 I'm getting the same results for division. Minus and plus give minus; minus and minus give plus. The rules must be the same for multiplication and division.

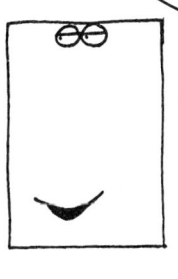 That's quite right. Multiplication and division follow the same rules. They are like opposite sides of a coin. Lets summarise the rules for division with negative numbers.

(a positive number) ÷ (a positive number) = (a positive number)

((+) ÷ (+) = (+))

(a positive number) ÷ (a negative number) = (a negative number)

((+) ÷ (−) = (−))

(a negative number) ÷ (a positive number) = (a negative number)

((−) ÷ (+) = (−))

(a negative number) ÷ (a negative number) = (a positive number)

((−) ÷ (−) = (+))

Brackets

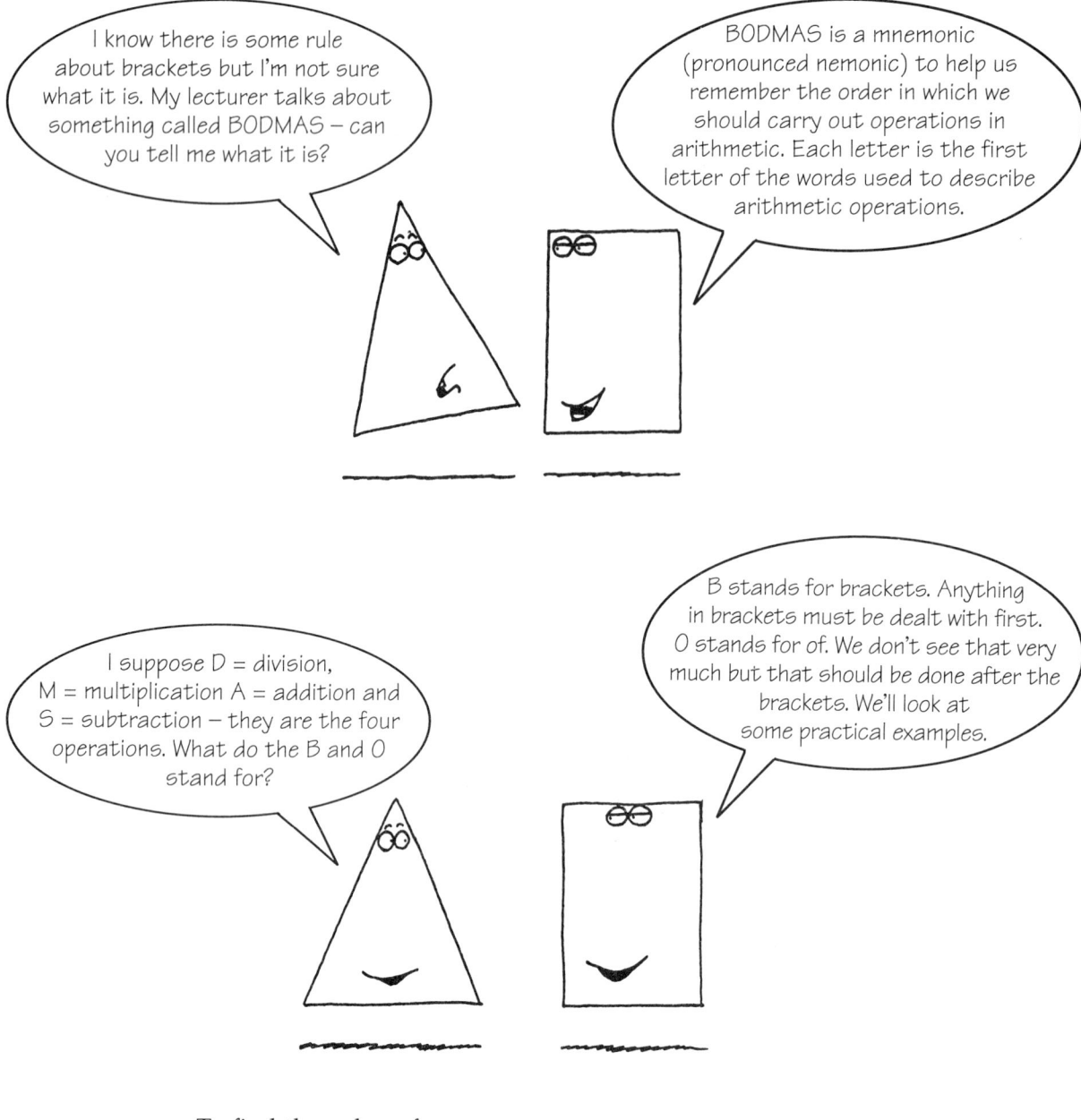

To find the value of

27 − (28 − 6)

we work out the sum in brackets first.

$$28 - 6 = 22$$
$$\text{Then } 27 - 22 = 5$$

Sometimes we have more than one set of brackets.

$$(3 \times 4) \div (2 \times 3)$$
$$3 \times 4 = 12 \quad 2 \times 3 = 6$$
$$12 \div 6 = 2$$

Until you are confident that you remember exactly what to do, it may help to write the letters over the sum to remind you what to do in which order.

Let us try
$$(3 + 10) \times (6 + 4) \div (2 \times 1) - 10$$

The first step is to identify the operations.

$$\overset{B}{(3 + 10)} \overset{M}{\times} \overset{B}{(6 + 4)} \overset{D}{\div} \overset{B}{(2 \times 1)} \overset{S}{-} 10$$

We have three sets of brackets so work these out first.
$$3 + 10 = 13 \quad 6 + 4 = 10 \quad 2 \times 1 = 2$$

$$\overset{M}{13} \overset{}{\times} \overset{D}{10} \overset{}{\div} \overset{S}{2} \overset{}{-} 10$$

Now we are left with multiplication, division and subtraction.

BO<u>D</u>M A<u>S</u> reminds us that division is first, followed by multiplication and finally subtraction.

$$13 \times 10 \div 2 - 10$$
$$10 \div 2 = 5$$
$$13 \times 5 = 65$$
$$65 - 10 = 55$$

So, the answer is 55.

Brackets are very important because they avoid misunderstanding.

$$\overset{A}{3} + \overset{M}{2 \times 5} = 13$$

$$2 \times 5 = 10 \quad 10 + 3 = 13 \text{ (Multiplication before addition)}$$

If we want to work out the addition first we must use brackets.

$$(3 + 2) \times 5 = 25$$
$$3 + 2 = 5 \quad 5 \times 5 = 25$$

It's very important to use your bracket keys when working out sums on your calculator.

[(....] Open bracket.
[....)] Close bracket.

Let's try a simple example.

Calculate 5 − (2 + 2)

| 5 | − | (| 2 | + | 2 |) | = | 1

Now see what happens if you leave out the brackets!

The same rules apply when you have more than one set of brackets.

$$(5 + 5) \times (6 - 4)$$

Remember BODMAS! Any calculation inside the bracket must be done before the rest.

| (| 5 | + | 5 |) | × | (| 6 | − | 4 |) | = | 20

Now see what happens when you leave out the brackets. Your answer will tell you how your calculator works.

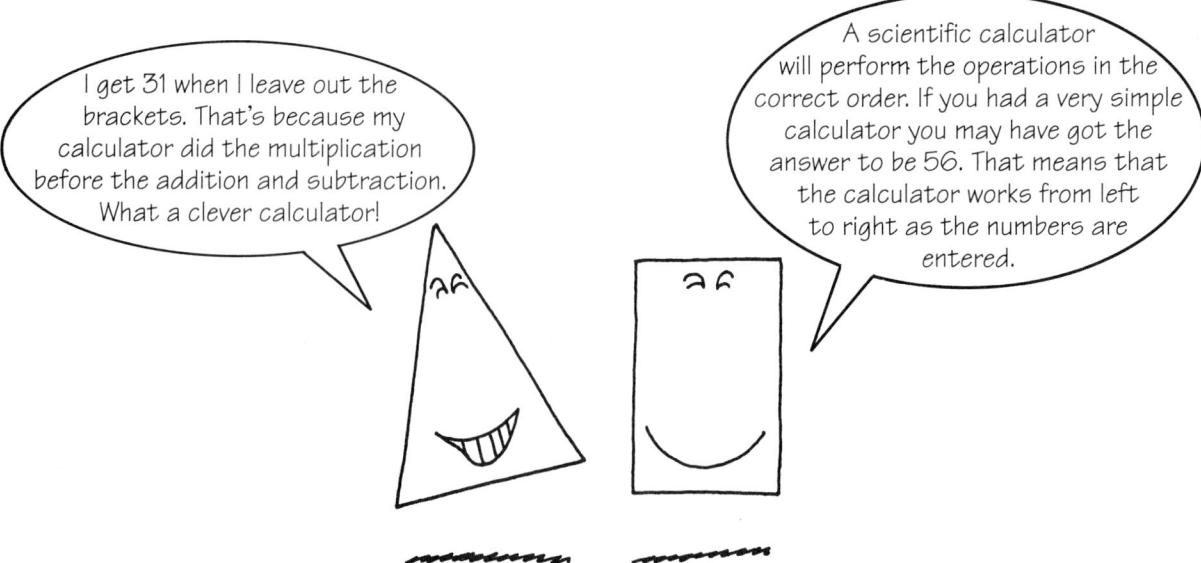

Try some for yourself

15. These examples should show you how important brackets are in avoiding ambiguity.

(a) $(5 + 3) - 4$

(b) $8 \times (1 - 6)$

(c) $5 - (3 - 8)$

(d) $(7 - 9) \times (5 \times 4)$

(e) $((-2) \div 7) \div 5$

(f) $((-3) - (-4)) \times 5$

(g) $(2 + (-2)) + (6 \times (-2))$

Brackets within brackets

Some calculators, and all scientific ones, can handle brackets within brackets. To check your own calculator try entering this calculation which contains 5 sets of brackets.

$$1 + (1 + (1 + (1 + (1 + (1 + 1))))) = 7$$

My calculator had no trouble with the first three open brackets then I got an E on the display. What does mean?

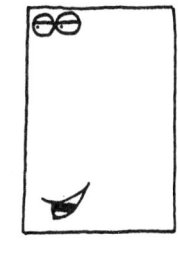

That tells you the number of brackets your calculator can cope with. You will need to remember that if you ever have a problem with more than 3 nested brackets.

That's all very interesting but not really useful – I can't see me having problems with lots of brackets. I want to know what I'm doing wrong when I try to work out 2 (4+6) on my calculator. I don't get the same answer as I get when I work it out on paper?

I think I can guess what the problem is. Can you just show me how you would put that into your calculator?

When I do it the long way I get 20 for my answer. My calculator is usually right but I'm not sure I believe it this time.

Your calculator is reading the entry as 24 + 6 so it gets the answer 30. 2 (4 + 6) is an abbreviation for 2 x (4 + 6). We mathematicians usually leave out the x sign before a bracket. You need to remember that. Try it again with the x sign included.

$2 \times (4 + 6) =$ 20

Let's summarise what we have discovered about arithmetic operations and calculators.

- **B**rackets
  Of
  Division
  Multiplication
  Addition
  Subtraction

  BODMAS is a mnemonic which helps us to remember the order in which we carry out arithmetic operations.

- ( )
  Open   Close

  Calculators have two bracket keys which are *always* used together.

- Calculators have limits on the number of sets of brackets they can handle.

- Not all calculators do multiplication and division before addition and subtraction.

- Calculators do not understand that the × sign is implicit in an expression like 3 (2 + 4). We must put it in when using a calculator.

What have you learnt about directed numbers?

Directed numbers are not really frightening now I have worked out the rules. My calculator is a great help.

- ✓ I can show positive and negative numbers as points on the number line.
- ✓ I can add, subtract, multiply and divide negative numbers.
- ✓ I can use the $+/-$ key on my calculator when I want to enter a negative number.
- ✓ I know what BODMAS stands for and why it is important.
- ✓ I can use the bracket keys on my calculator.

Further Questions

1. Without using a calculator work out each of the following:

(a) $5 + 8$

(b) $3 - 7$

(c) $8 - 15$

(d) $(-3) + 6$

(e) $(-4) - 8$

(f) $3 + (-1)$

(g) $15 + (-16)$

(h) $(-5) + (-8)$

(i) $(-15) + (-5)$

(j) $(-3) - (-4)$

(k) $(-6) - (-10)$

(l) $(-15) - (-16)$

(m) $(-3) + (-4)$

(n) $16 - (-15)$

(o) $9 - (-10)$

(p) $(-8) - (-8)$

(q) $1 + 32$

(r) $11 - (-10)$

2. Without a calculator work out each of the following:
(a) $4 \times (-5)$
(b) $(-5) \times 4$
(c) $(-10(\times (-2)$
(d) $(-1) \times (-20)$
(e) $(-6) \times 3$
(f) $10 \div (-5)$
(g) $(-5) \div 10$
(h) $(-10) \div (-2)$
(i) $(-20) \div (-4)$
(j) $(-6) \div 3$

3. Without a calculator work out each of the following:
(a) $3(4-5)$
(b) $6+(7-3)$
(c) $(7+4)-5$
(d) $(7+3) \div 5$
(e) $(7-4) \div 3$
(f) $3 \times (4+5)$
(g) $10(5-4)$
(h) $6(3-(-5))$
(i) $(5-3) \times (4+4)$
(j) $-(5-2(3+1))$

4. Use a calculator to work out each of the following:
(a) $1547 + (-2463)$
(b) $3004 + (-2989)$
(c) $(-1672) \times (-1534)$
(d) $(4587) \div (-1529)$
(e) $2845 \div (-569)$
(f) $(1936 - 724) \times (15 - 164)$
(g) $365 \times (298 - (-122))$
(h) $890(635 - (189 - 259))$

Solutions

1. (a)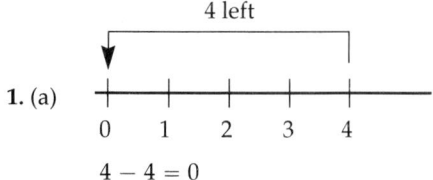
$4 - 4 = 0$

(b)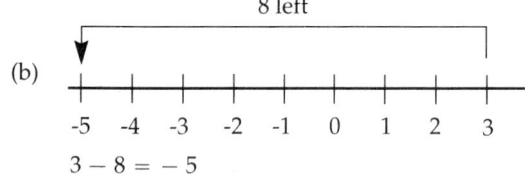
$3 - 8 = -5$

(c)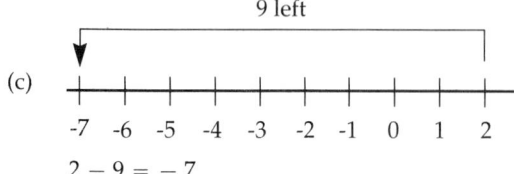
$2 - 9 = -7$

(d)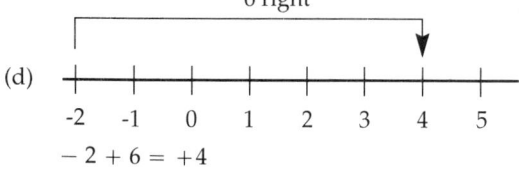
$-2 + 6 = +4$

2. (a) $\boxed{4}\ \boxed{+/-}\quad -4$
(b) $\boxed{6}\ \boxed{+/-}\quad -6$
(c) $\boxed{1}\boxed{7}\boxed{8}\boxed{2}\ \boxed{+/-}\quad -1782$
(d) $\boxed{2}\boxed{9}\boxed{4}\boxed{1}\ \boxed{+}\quad -2941$

3. (a)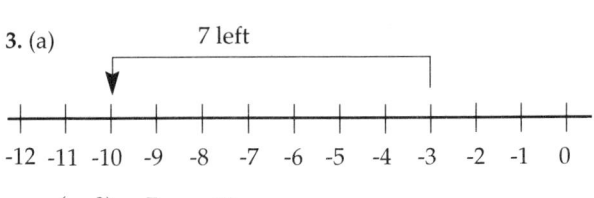
$(-3) - 7 = -10$

(b)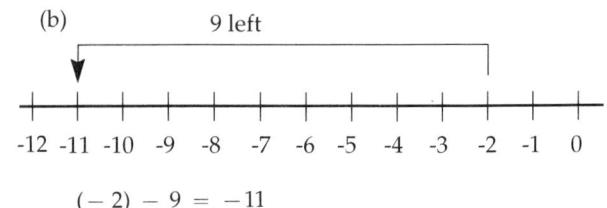
$(-2) - 9 = -11$

(c)
```
6 left
```
$-5\ -4\ -3\ -2\ -1\ 0\ 1\ 2$

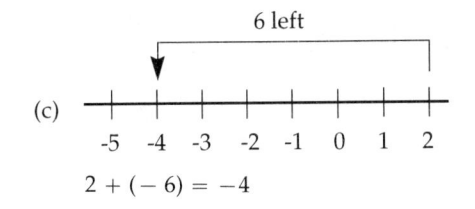

$2 + (-6) = -4$

(d)
```
7 left
```
$-3\ -2\ -1\ 0\ 1\ 2\ 3\ 4$

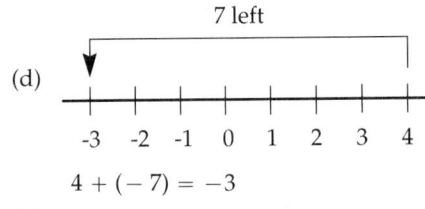

$4 + (-7) = -3$

(e)
```
8 left
```
$-11\ -10\ -9\ -8\ -7\ -6\ -5\ -4\ -3\ -2\ -1\ 0$

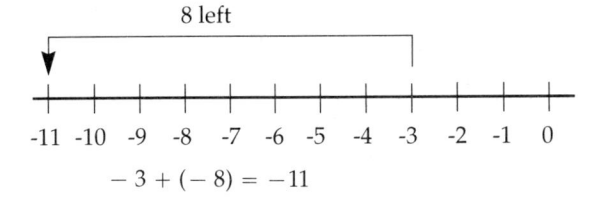

$-3 + (-8) = -11$

(f)
```
2 left
```
$-4\ -3\ -2\ -1\ 0$

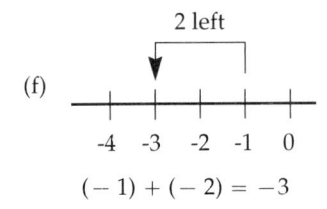

$(-1) + (-2) = -3$

4. (a) (i) $\boxed{3}\ \boxed{+}\ \boxed{4}\ \boxed{+/-}\ \boxed{=}$ -1
(ii) $\boxed{3}\ \boxed{-}\ \boxed{4}\ \boxed{=}$ -1
(b) (i) $\boxed{3}\ \boxed{+}\ \boxed{1}\ \boxed{0}\ \boxed{+/-}\ \boxed{=}$ -7
(ii) $\boxed{3}\ \boxed{-}\ \boxed{1}\ \boxed{0}\ \boxed{=}$ -7
(c) (i) $\boxed{2}\ \boxed{+}\ \boxed{7}\ \boxed{+/-}\ \boxed{=}$ -5
(ii) $\boxed{2}\ \boxed{-}\ \boxed{7}\ \boxed{=}$ -5
(d) (i) $\boxed{5}\ \boxed{+}\ \boxed{8}\ \boxed{+/-}\ \boxed{=}$ -3
(ii) $\boxed{5}\ \boxed{-}\ \boxed{8}\ \boxed{=}$ -3

5. (a) (i) $\boxed{2}\ \boxed{+/-}\ \boxed{+}\ \boxed{7}\ \boxed{+/-}\ \boxed{=}$ -9
(ii) $\boxed{2}\ \boxed{+/-}\ \boxed{-}\ \boxed{7}\ \boxed{=}$ -9
(b) (i) $\boxed{3}\ \boxed{+/-}\ \boxed{+}\ \boxed{1}\ \boxed{1}\ \boxed{+/-}\ \boxed{=}$ -14
(ii) $\boxed{3}\ \boxed{+/-}\ \boxed{-}\ \boxed{1}\ \boxed{1}\ \boxed{=}$ -14

6. (a) $\boxed{2}\ \boxed{-}\ \boxed{3}\ \boxed{+/-}\ \boxed{=}$ 5
(b) $\boxed{4}\ \boxed{-}\ \boxed{7}\ \boxed{+/-}\ \boxed{=}$ 11
(c) $\boxed{6}\ \boxed{-}\ \boxed{2}\ \boxed{+/-}\ \boxed{=}$ 8
(d) $\boxed{8}\ \boxed{-}\ \boxed{1}\ \boxed{0}\ \boxed{+/-}\ \boxed{=}$ 18
(e) $\boxed{1}\ \boxed{-}\ \boxed{1}\ \boxed{+/-}\ \boxed{=}$ 2

7. $-£36 + £36 = £0;\ £36$

8. $-53 + 73 = 20;\ 73$ metres

9. $-3 + (-7) = -10;\ -7°C$

10. $£13 - £25 = -£12$
$-£12 - £20 = -£32$
£32 overdrawn

11. (a) (i) $\boxed{2}\ \boxed{\times}\ \boxed{3}\ \boxed{+/-}\ \boxed{=}$ -6
(ii) $2 \times 3 = 6$
(b) (i) $\boxed{4}\ \boxed{\times}\ \boxed{7}\ \boxed{+/-}\ \boxed{=}$ -28
(ii) $4 \times 7 = 28$
(c) (i) $\boxed{6}\ \boxed{\times}\ \boxed{1}\ \boxed{0}\ \boxed{+/-}\ \boxed{=}$ -60
(ii) $6 \times 10 = 60$
(d) (i) $\boxed{1}\ \boxed{\times}\ \boxed{1}\ \boxed{+/-}\ \boxed{=}$ -1
(ii) $1 \times 1 = 1$

In part (i) the answer is negative.
In part (ii) the answer is positive.

12. (a) $\boxed{2}\ \boxed{+/-}\ \boxed{\times}\ \boxed{3}\ \boxed{=}$ -6
(b) $\boxed{4}\ \boxed{+/-}\ \boxed{\times}\ \boxed{7}\ \boxed{=}$ -28
(c) $\boxed{6}\ \boxed{+/-}\ \boxed{\times}\ \boxed{1}\ \boxed{0}\ \boxed{=}$ -60
(d) $\boxed{1}\ \boxed{+/-}\ \boxed{\times}\ \boxed{1}\ \boxed{=}$ -1

13. (a) $\boxed{2}\ \boxed{+/-}\ \boxed{\times}\ \boxed{3}\ \boxed{+/-}\ \boxed{=}$ 6
(b) $\boxed{4}\ \boxed{+/-}\ \boxed{\times}\ \boxed{7}\ \boxed{+/-}\ \boxed{=}$ 28
(c) $\boxed{6}\ \boxed{+/-}\ \boxed{\times}\ \boxed{1}\ \boxed{0}\ \boxed{+/-}\ \boxed{=}$ 60
(d) $\boxed{1}\ \boxed{+/-}\ \boxed{\times}\ \boxed{1}\ \boxed{+/-}\ \boxed{=}$ 1

14. (a) $\boxed{2}\ \boxed{0}\ \boxed{\div}\ \boxed{4}\ \boxed{+/-}\ \boxed{=}$ -5
(b) $\boxed{3}\ \boxed{6}\ \boxed{+/-}\ \boxed{\div}\ \boxed{1}\ \boxed{2}\ \boxed{+/-}\ \boxed{=}$ 3
(c) $\boxed{4}\ \boxed{2}\ \boxed{+/-}\ \boxed{\div}\ \boxed{1}\ \boxed{4}\ \boxed{=}$ -3
(d) $\boxed{8}\ \boxed{1}\ \boxed{\div}\ \boxed{9}\ \boxed{+/-}\ \boxed{=}$ -9
(e) $\boxed{1}\ \boxed{5}\ \boxed{+/-}\ \boxed{\div}\ \boxed{3}\ \boxed{=}$ -5
(f) $\boxed{2}\ \boxed{7}\ \boxed{+/-}\ \boxed{\div}\ \boxed{3}\ \boxed{+/-}\ \boxed{=}$ 9
(g) $\boxed{3}\ \boxed{9}\ \boxed{+/-}\ \boxed{\div}\ \boxed{1}\ \boxed{3}\ \boxed{+/-}\ \boxed{=}$ 3
(h) $\boxed{5}\ \boxed{0}\ \boxed{4}\ \boxed{+/-}\ \boxed{\div}\ \boxed{5}\ \boxed{6}\ \boxed{=}$ -9
(i) $\boxed{1}\ \boxed{2}\ \boxed{\div}\ \boxed{4}\ \boxed{=}$ 3

15. (a) 4
(b) -40
(c) 10
(d) -18
(e) -0.0057
(f) 5
(g) -12

31

3 Fractions

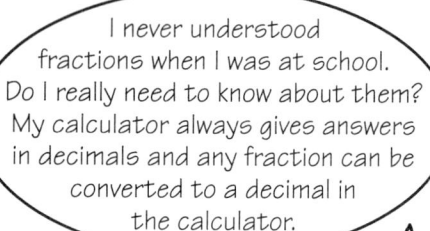

I never understood fractions when I was at school. Do I really need to know about them? My calculator always gives answers in decimals and any fraction can be converted to a decimal in the calculator.

You may think that in this age of calculators and computers, fractions are out-of-date. There are times when it is still necessary to use fractions, I'm afraid.

What is a fraction?

A fraction is part of a whole. A clock is marked off in hours but time is still measured in fractions of an hour.

1.15
quarter past one

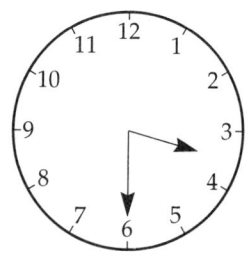

3.30
half-past three

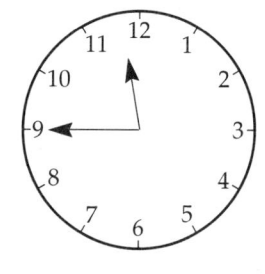

11.45
quarter to twelve

Tape measures are marked in centimetres but we still ask for half a metre rather than 50 cm.

Screws are made in both metric and fractional sizes, for example, $\frac{5}{8}"$, $1\frac{3}{8}"$.

Cake tins have their diameters marked in inches: $7\frac{1}{2}"$, $8"$, $9\frac{1}{2}"$.

Hat sizes for men are also measured in fractions: $6\frac{7}{8}"$, $7\frac{1}{2}"$, etc.

You may find the following illustrations helpful when thinking of fractions.

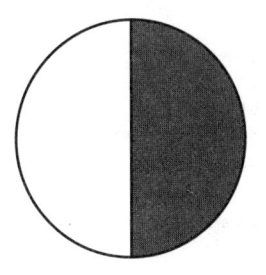

$\frac{1}{2}$ of the circle $\frac{1}{4}$ of the circle $\frac{1}{8}$ of the circle

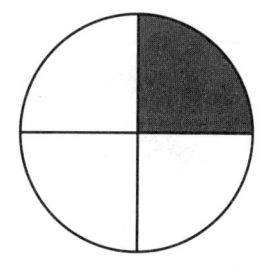

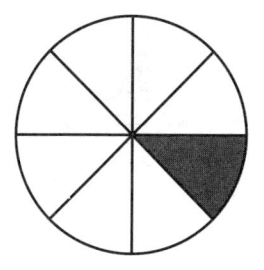

Each of the circles is divided into a number of equal pieces.

The first circle is divided in 2 equal pieces.
Each piece is $\frac{1}{2}$ of the circle.

The second circle is divided into 4 equal pieces.
Each piece is $\frac{1}{4}$ of the circle.

The third circle is divided into 8 equal pieces.
Each piece is $\frac{1}{8}$ of the circle.

We can divide any regular shape into a number of equal parts.

This chocolate wafer biscuit bar is marked into 4 fingers – each finger is $\frac{1}{4}$ of the bar.

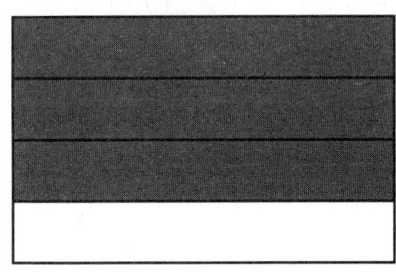

3 fingers = 3 parts of the bar = $\frac{3}{4}$ of the bar.

$$\frac{3}{4} = \frac{\text{number of shaded parts}}{\text{total number of parts}} = \frac{\text{numerator}}{\text{denominator}}$$

numerator = number on the top of a fraction.
denominator = number on the bottom of a fraction.

Try some for yourself

1. Look at the following chocolate bars and write down the fraction shown by the shaded part.

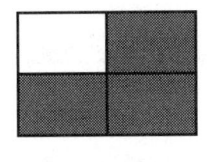

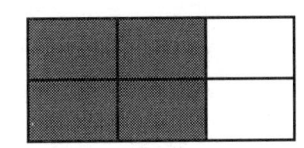

(a)   (b)   (c)

2. Look at these pizzas and write down the fraction indicated by the shaded parts.

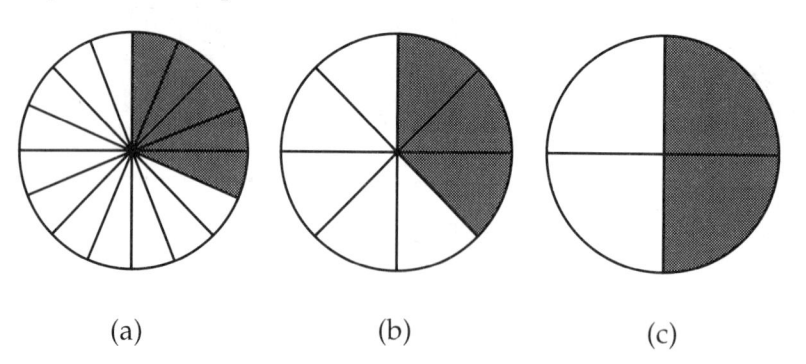

(a)   (b)   (c)

I can see how to identify fractions when it's just parts of a whole. My problem is when it's a little more complicated. I never knew what to do when I was asked to write a fraction in its simplest form – cancelling I think it was called. That was a total mystery.

To cancel fractions you need to understand about equivalent fractions. I think we'll start there and then find out how to reduce a fraction to its simplest form.

Equivalent fractions

$\frac{1}{2}$ $\frac{2}{4}$ $\frac{4}{8}$

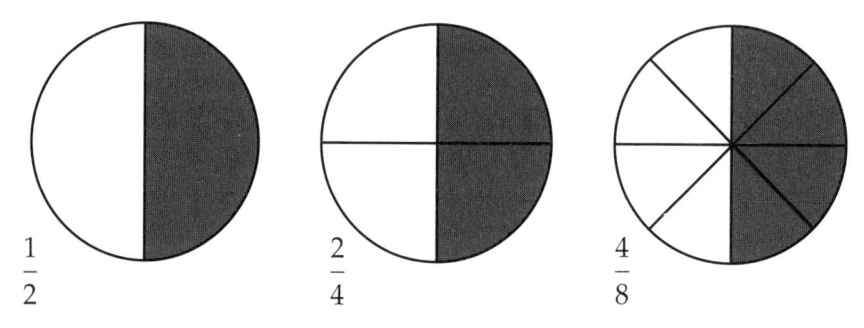

These are three different fractions, but the areas of the shaded parts are equal.

So
$$\frac{1}{2} = \frac{2}{4} = \frac{4}{8}$$

Fractions which are equal but have different numerators and denominators are called **equivalent fractions**.

To find an equivalent fraction for any given fraction we first have to multiply the numerator and denominator by the *same* whole number.

$$\frac{1}{2} \times \frac{2}{2} = \frac{2}{4} \qquad \frac{2}{4} \times \frac{2}{2} = \frac{4}{8}$$

Let's find some equivalent fractions for $\frac{1}{3}$,

$$\frac{1}{3} \times \frac{2}{2} = \frac{2}{6}, \qquad \frac{1}{3} = \frac{2}{6}$$

$$\frac{1}{3} \times \frac{3}{3} = \frac{3}{9}, \qquad \frac{1}{3} = \frac{3}{9}$$

$$\frac{1}{3} \times \frac{4}{4} = \frac{4}{12}, \qquad \frac{1}{3} = \frac{4}{12}$$

So
$$\frac{2}{6} = \frac{3}{9} = \frac{4}{12} = \frac{1}{3}$$

These are all equivalent fractions.

That looks quite simple but you still haven't told me how to cancel. How is cancelling connected to equivalent fractions?

*So far, we have found equivalent fractions by multiplying the numerator and denominator by the same number. If we **divide** the numerator and denominator by the same number we can also find equivalent fractions. This is sometimes called **cancelling**.*

Cancelling

Let's find some equivalent fractions for $\frac{24}{32}$.

$$\frac{24}{32} \div \frac{4}{4} = \frac{6}{8} \qquad \frac{6}{8} \div \frac{2}{2} = \frac{3}{4}$$

35

$\frac{3}{4}$ is the simplest fraction equivalent to $\frac{24}{32}$ because there is no whole number which divides both 3 and 4. We have reduced the fraction $\frac{24}{32}$ to its lowest (or simplest) form.

A number which divides a given number an exact number of times is called a **factor** of that number. A number which divides two different numbers exactly is called a **common factor**. It is a factor which is common to both numbers.

> So if I want to reduce a fraction to its simplest form, I need to look for common factors of the numerator and denominator.

> That's exactly what you need to do. It's often best to start with the prime numbers like 2, 3, and 5. Any even number can be divided by 2.

Let's try to reduce $\frac{48}{60}$ to its simplest form.

48 and 60 are even numbers so we'll divide by 2.

$\frac{48 \div 2}{60 \div 2} = \frac{24}{30}$ 24 and 30 are even numbers so divide by 2 again.

$\frac{24 \div 2}{30 \div 2} = \frac{12}{15}$ 12 is even but 15 is odd so 2 is no use. Let's try the next prime number, 3.

$\frac{12 \div 3}{15 \div 3} = \frac{4}{5}$ 4 is even and 5 is odd. There are no more common factors.

$\frac{4}{5}$ is the simplest form of $\frac{48}{60}$.

▽ Try some for yourself

3. Which of the following are equivalent fractions for $\frac{2}{5}$?
   (a) $\frac{4}{5}$ (b) $\frac{4}{10}$ (c) $\frac{4}{15}$ (d) $\frac{10}{25}$

4. Which of the following are pairs of equivalent fractions?
   (a) $\frac{2}{7}, \frac{4}{28}$ (b) $\frac{3}{16}, \frac{6}{8}$ (c) $\frac{4}{6}, \frac{2}{3}$

5. Reduce each of the following fractions to its simplest form.
   (a) $\frac{15}{30}$ (b) $\frac{15}{60}$ (c) $\frac{25}{100}$ (d) $\frac{54}{70}$

Adding fractions

That makes cancelling much clearer. I had no idea it was so simple. It always seemed like magic when the teacher did it! Now can you show me how to add and subtract fractions?

Now you understand about equivalent fractions, adding and subtracting fractions should make more sense to you.

To add two fractions with the same denominator is quite straightforward; you just add the numerators.

$$\frac{3}{8} + \frac{2}{8} = \frac{3+2}{8} = \frac{5}{8}$$

If the denominators are different, then addition is a bit more complicated.

The first thing you must do is to make the denominators the same. This is called **finding the common denominator**. We need to use equivalent fractions.

37

Let us try $\quad \dfrac{1}{4} + \dfrac{3}{8}$

We want to make the denominators the same.

We can't change $\dfrac{3}{8}$ into $\dfrac{?}{4}$ because 3 and 8 have no common factors.

Let's see if we can change $\dfrac{1}{4}$ into $\dfrac{?}{8}$

$$\dfrac{1}{4} \times \dfrac{2}{2} = \dfrac{2}{8} \quad \text{so} \quad \dfrac{1}{4} = \dfrac{2}{8}$$

Now both fractions have a denominator of 8.

$$\dfrac{2}{8} + \dfrac{3}{8} = \dfrac{2+3}{8} = \dfrac{5}{8}$$

8 is the **common denominator** of 4 and 8 so we can add the fractions.

▽ Try some for yourself

Evaluate (that's the fancy word mathematicians use when they mean 'work out') these and give the answer in its simplest form.

6. (a) $\dfrac{3}{10} + \dfrac{5}{10}$ (c) $\dfrac{8}{9} + \dfrac{1}{5}$

(b) $\dfrac{1}{3} + \dfrac{1}{6}$

Subtracting fractions

We found that adding fractions was easy when they had the same denominators. The same thing is true when we subtract fractions with the same denominator.

$$\dfrac{6}{9} - \dfrac{2}{9} = \dfrac{6-2}{9} = \dfrac{4}{9}$$

$$\dfrac{5}{12} - \dfrac{3}{12} = \dfrac{5-3}{12} = \dfrac{2}{12} \div \dfrac{2}{2} = \dfrac{1}{6}$$

If the denominators are different then subtraction is more complicated. We must first find the common denominator.

As with addition, the first step is to make the denominators the same.

It's not always easy to find the common denominator. Sometimes it's obvious; occasionally it's a question of trial and error.

If you are having difficulty finding the common denominator you can just multiply the denominators together. This *always* gives a common denominator but it does mean that numbers can be rather large at times.

Let's try a subtraction.

$$\frac{1}{2} - \frac{1}{4}$$

Here we can see that 4 will be the common denominator, $2 \times 2 = 4$.

$$\frac{1}{2} \times \frac{2}{2} = \frac{2}{4}$$

$$\frac{2}{4} - \frac{1}{4} = \frac{2-1}{4} = \frac{1}{4}$$

Now we'll try an example where the common denominator is not so obvious.

$$\frac{9}{16} - \frac{2}{5}$$

Here we shall have to multiply the two denominators to find a common denominator.

$$16 \times 5 = 80$$

We multiply 16×5 to get 80, so we must multiply 9 by 5.

$$\frac{9}{16} \times \frac{5}{5} = \frac{45}{80}$$

We multiply 5×16 to get 80, so we must multiply 2×16.

$$\frac{2}{5} \times \frac{16}{16} = \frac{32}{80}$$

$$\frac{45}{80} - \frac{32}{80} = \frac{45-32}{80} = \frac{13}{80}$$

▽ Try some for yourself

7. (a) $\dfrac{9}{10} - \dfrac{5}{10}$ (d) $\dfrac{5}{8} - \dfrac{2}{7}$

(b) $\dfrac{11}{32} - \dfrac{7}{32}$ (e) $\dfrac{11}{12} - \dfrac{4}{9}$

(c) $\dfrac{7}{16} - \dfrac{2}{16}$ (f) $\dfrac{5}{16} + \dfrac{1}{3} - \dfrac{1}{6}$

Multiplying fractions

We've looked at adding and subtracting fractions. Can you show me how to multiply and divide fractions? I think I know how to multiply but dividing is a bit peculiar isn't it?

Multiplying fractions is quite straightforward. You multiply the numerators together and multiply the denominators together.

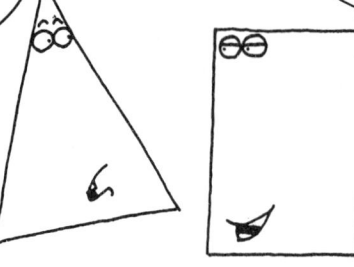

What are we actually doing when we multiply?

When you multiply by a whole number you are really just adding a number a specific number of times.

$$4 \times 3 = \underbrace{4 + 4 + 4}_{3 \text{ times}} = 12$$

This is called **repeated addition** and was the earliest method of multiplication.

When you multiply a fraction by a whole number you follow the same procedure.

$$\frac{3}{5} \times 3 = \underbrace{\frac{3}{5} + \frac{3}{5} + \frac{3}{5}}_{3 \text{ times}} = \frac{9}{5}$$

What happens when we multiply?

$$\frac{3}{5} \times 3 = \frac{3 \times 3}{5} = \frac{9}{5}$$

We get exactly the same answer!

In fact, *any* two fractions can be multiplied together by multiplying the numerators and multiplying the denominators.

$$\frac{3}{4} \times \frac{2}{3} = \frac{\overbrace{3 \times 2}^{\text{numerators}}}{\underbrace{4 \times 3}_{\text{denominators}}} = \frac{6}{12}$$

$$\frac{2}{7} \times \frac{3}{5} = \frac{2 \times 3}{7 \times 5} = \frac{6}{35}$$

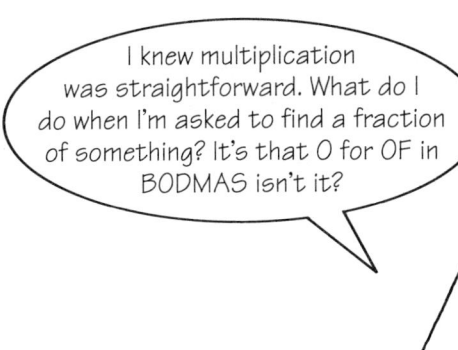

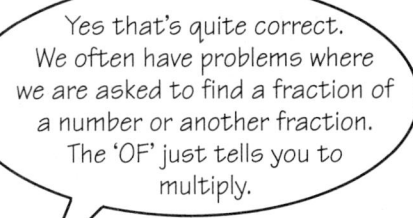

> I knew multiplication was straightforward. What do I do when I'm asked to find a fraction of something? It's that O for OF in BODMAS isn't it?

> Yes that's quite correct. We often have problems where we are asked to find a fraction of a number or another fraction. The 'OF' just tells you to multiply.

$$\frac{1}{2} \text{ of } 16 = 16 \times \frac{1}{2} = \frac{16}{2} = 8$$

$$\frac{3}{4} \text{ of } \frac{1}{2} = \frac{1}{2} \times \frac{3}{4} = \frac{3}{8}$$

Sometimes, when we multiply two or more fractions we find the answer contains large numbers. There is a quick way to multiply fractions which makes the multiplication easier.

$$\frac{6}{15} \times \frac{5}{9} = \frac{6 \times 5}{15 \times 9} = \frac{30}{135}$$

Multiplying 15 by 9 is not easy. We can make it easier by looking for some common factors of both numerators and denominators.

$$15 = 3 \times 5 \qquad 6 = 2 \times 3 \qquad 9 = 3 \times 3$$

$$\frac{\cancel{6}^2}{\cancel{15}_3} \times \frac{\cancel{5}^1}{\cancel{9}_3} = \frac{2}{9}$$

This process of simplifying numbers first by finding common factors is called **cancelling**.

Let's look at another problem.

$$\frac{4}{49} \times \frac{7}{16}$$

4 is a factor common to 4 and 16 (16 = 4 × 4)

7 is a factor common to 7 and 49 (49 = 7 × 7)

Cancelling,

$$\frac{\cancel{4}^1}{\cancel{49}_7} \times \frac{\cancel{7}^1}{\cancel{16}_4} = \frac{1 \times 1}{7 \times 4} = \frac{1}{28}$$

This is the answer in its **simplest (lowest)** form.

▽ Try some for yourself

8. Evaluate the following. In each case give the answer in its simplest form.

(a) $\dfrac{11}{12} \times \dfrac{3}{11}$

(b) $\dfrac{5}{6} \times \dfrac{9}{10}$

(c) $\dfrac{17}{33} \times \dfrac{12}{34} \times \dfrac{3}{4}$

(d) Find $\dfrac{3}{4}$ of 12

(e) Find $\dfrac{4}{5}$ of $\dfrac{1}{2}$

(f) A school contains 720 students. $\frac{5}{8}$ are male and $\frac{3}{8}$ are female. Give the number of males and the number of females.

Dividing fractions

Now we come to the real mystery. How do I divide by a fraction? I remember something about turning upside down and multiplying, Why?

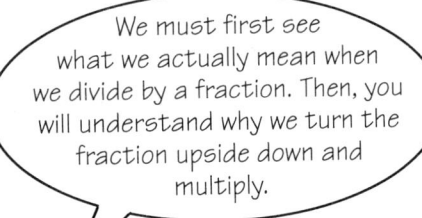

We must first see what we actually mean when we divide by a fraction. Then, you will understand why we turn the fraction upside down and multiply.

What do we mean when we write

$$6 \div \tfrac{1}{3}?$$

Do you expect the answer to be bigger or smaller than 6?

What happens when we divide whole numbers?

$6 \div 3$ asks for the number of threes in six.

$6 \div 3 = 2$ because $2 \times 3 = 6$

In the same way, $6 \div \tfrac{1}{3}$ is asking for the number of thirds ($\tfrac{1}{3}$) in 6. Think of 6 as 6 whole rectangles.

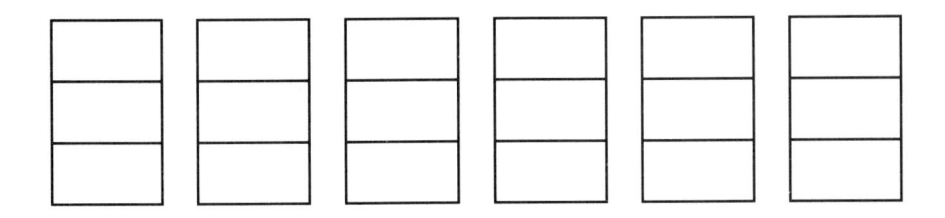

Each rectangle contains three third-rectangles, and 6 rectangles contain 6×3 third-rectangles.

$$1 = 3 \times \tfrac{1}{3} \qquad 6 = 6 \times 1 = 6 \times 3\tfrac{1}{3}$$

$$\text{So } 6 \div \tfrac{1}{3} = 6 \times 3 = 18$$

What happens when we divide one fraction by another fraction?

What is the meaning of $\tfrac{1}{2} \div \tfrac{1}{4}$?

We want to know how many $\tfrac{1}{4}$ are to be found in $\tfrac{1}{2}$.

Think of half a rectangle.

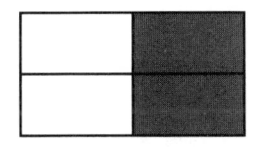

Each whole rectangle contains 4 quarter-rectangles.

So $\tfrac{1}{2}$ a whole rectangle contains $\tfrac{1}{2} \times 4$ quarter-rectangles.

Half a rectangle contains $2 \times \tfrac{1}{4}$ rectangles.

$$\text{So } \tfrac{1}{2} \div \tfrac{1}{4} = 2.$$

$$\text{In fact } \tfrac{1}{2} \div \tfrac{1}{4} = \tfrac{1}{2} \times \tfrac{4}{1} = 2$$

This suggests a rule for dividing by a fraction.

> To divide by a fraction turn the fraction upside down and multiply.

Let's try out the rule.

Evaluate $\frac{1}{3} \div \frac{1}{6}$

Using our rule

$$\frac{1}{3} \div \frac{1}{6} = \frac{1}{3} \times \frac{6}{1} = 2$$

As we expect, there are two $\frac{1}{6}$ in each $\frac{1}{3}$.

Turning the fraction upside down is called finding the **reciprocal**.

The reciprocal of 6 is $\frac{1}{6}$.

The reciprocal of $\frac{5}{6}$ is $\frac{6}{5}$.

How do you find the reciprocal? How do you know that the reciprocal of $\frac{5}{6}$ is $\frac{6}{5}$? I don't understand.

Look at the numbers and their reciprocals. Multiply each number by its reciprocal and see what happens.

$$6 \times \frac{1}{6} = \frac{6}{6} = 1$$
$$\frac{5}{6} \times \frac{6}{5} = \frac{30}{30} = 1$$

When I multiply a number by its reciprocal the answer is one. Is that always true?

Yes, when a number multiplied by its reciprocal the answer is always 1.

Try some for yourself

9. (a) $10 \div \frac{1}{2}$ (c) $\frac{16}{9} \div \frac{4}{3}$

(b) $\frac{2}{5} \div \frac{5}{7}$ (d) $\frac{14}{8} \div \frac{7}{16}$

Manipulating mixed numbers

I think I now know how to add, subtract, multiply and divide fractions. I notice that you have only looked at simple fractions – what happens when I have whole numbers and fractions mixed together?

Those are called mixed numbers because whole numbers and fractions are mixed together. Mixed numbers can be turned into improper fractions – that means that the numerator is bigger than the denominator. Proper fractions have the denominator larger than the numerator.

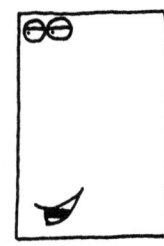

The first step in handling mixed numbers is to learn how to change them to improper fractions.

Let's write $2\frac{3}{4}$ as an improper (top-heavy) fraction.

We first look at the whole number 2.

$$2 = 2 \times 1$$

How many $\frac{1}{4}$ are there in the whole number 1?

There are $4 \times \frac{1}{4}$ in 1 so

$$2 = 2 \times \frac{4}{4} = \frac{8}{4}$$

There are $8 \times \frac{1}{4}$ in 2.

We can write the whole number 2 as $\frac{8}{4}$.

Now we first add on the $\frac{3}{4}$.

$$2\frac{3}{4} = \frac{8}{4} + \frac{3}{4} = \frac{11}{4}$$

So $2\frac{3}{4}$ is the improper fraction $\frac{11}{4}$.

If we reverse the process, we can change an improper fraction into a mixed number.

$\frac{11}{4}$ is another way of writing $11 \div 4$.

$11 \div 4 = 2$ with remainder 3. $(2 \times 4 = 8 \quad 11 - 8 = 3)$

So $\frac{11}{4} = 2\frac{3}{4}$.

▽ Try some for yourself

10. Convert each of these to improper fractions.

   (a) $3\frac{5}{6}$ (b) $6\frac{1}{4}$ (c) $2\frac{28}{33}$ (d) $1\frac{9}{16}$

11. Convert each of these improper fractions to mixed numbers.

   (a) $\frac{23}{6}$ (b) $\frac{25}{4}$ (c) $\frac{84}{31}$ (d) $\frac{15}{4}$

Adding and subtracting mixed numbers

Work out $5\frac{3}{4} + 1\frac{1}{6}$

$$\begin{aligned}5\tfrac{3}{4} + 1\tfrac{1}{6} &= 5 + \tfrac{3}{4} + 1 + \tfrac{1}{6}\\ &= 5 + 1 + \tfrac{3}{4} + \tfrac{1}{6}\\ &= 6 + \tfrac{3}{4} + \tfrac{1}{6}\\ &= 6 + \tfrac{9}{12} + \tfrac{2}{12}\\ &= 6 + \tfrac{11}{12}\\ &= 6\tfrac{11}{12}\end{aligned}$$

> First add the whole number parts.

> Add the fractions by putting over a common denominator.

> Add the fraction to the whole number.

Work out $3\frac{5}{8} - 2\frac{1}{4}$

$$\begin{aligned}3\tfrac{5}{8} - 2\tfrac{1}{4} &= 3 + \tfrac{5}{8} - 2 - \tfrac{1}{4}\\ &= 3 - 2 + \tfrac{5}{8} - \tfrac{1}{4}\\ &= 1 + \tfrac{5}{8} - \tfrac{1}{4}\\ &= 1 + \tfrac{5}{8} - \tfrac{2}{8}\\ &= 1 + \tfrac{3}{8}\\ &= 1\tfrac{3}{8}\end{aligned}$$

> First subtract the whole number parts.

> Subtract the fractions by putting over a common denominator.

> Add the fraction to the whole number.

That's all quite easy. I remember most of that. I can remember getting into a muddle when the second fraction was bigger that the first. I just didn't know what to do.

That sort of problem is a bit tricky and often causes trouble. I'll show you one way to handle that situation.

$$\begin{aligned}2\tfrac{1}{2} - 1\tfrac{3}{4} &= 2 + \tfrac{1}{2} - 1 - \tfrac{3}{4}\\ &= 2 - 1 + \tfrac{1}{2} - \tfrac{3}{4}\\ &= 1 + \tfrac{1}{2} - \tfrac{3}{4}\\ &= 1 + \tfrac{2}{4} - \tfrac{3}{4}\\ &= 1 - \tfrac{1}{4}\\ &= \tfrac{4}{4} - \tfrac{1}{4}\\ &= \tfrac{3}{4}\end{aligned}$$

> Subtract the whole number parts.

> Subtract the fraction parts.

> Subtract the fraction from the whole number by converting the whole number to an improper fraction.

Try some for yourself

12. (a) $1\frac{5}{8} + 2\frac{1}{4}$ (c) $2\frac{1}{5} + 1\frac{3}{10}$ (e) $4\frac{1}{2} - 1\frac{3}{4}$

(b) $5\frac{1}{6} + \frac{3}{8}$ (d) $5\frac{3}{4} - 4\frac{1}{8}$ (f) $2\frac{1}{3} - 1\frac{1}{6}$

Multiplying and dividing mixed numbers

I think this is going to be very complicated. I expect I have to multiply or divide the whole numbers and then the fractions. I'll get so confused.

You are quite wrong. When we multiply or divide mixed numbers we change them to improper fractions before we start. Then it's just like multiplying or dividing fractions.

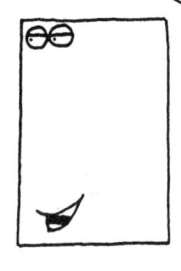

Try $1\frac{3}{4} \times 2\frac{1}{3}$

$1\frac{3}{4} = \frac{4}{4} + \frac{3}{4} = \frac{7}{4}$

$2\frac{1}{3} = \frac{6}{3} + \frac{1}{3} = \frac{7}{3}$

First change the mixed numbers to improper fractions.

$1\frac{3}{4} \times 2\frac{1}{3} = \frac{7}{4} \times \frac{7}{3} = \frac{49}{12}$

Multiply the fractions.

$\frac{49}{12} = 4\frac{1}{12}$ $\begin{pmatrix} 4 \times 12 = 48 \\ 49 - 48 = 1 \end{pmatrix}$

Change back to a mixed number.

Now try $1\frac{5}{8} \div 1\frac{1}{6}$

$1\frac{5}{8} = \frac{8}{8} + \frac{5}{8} = \frac{13}{8}$

$1\frac{1}{6} = \frac{6}{6} + \frac{1}{6} = \frac{7}{6}$

First change the mixed numbers to improper fractions.

$\dfrac{13}{8} \div \dfrac{7}{6} = \dfrac{13}{\cancel{8}_4} \times \dfrac{\cancel{6}^3}{7}$

$= \frac{39}{28}$

$= 1\frac{11}{28}$

Divide the fractions by turning the second one upside down and multiplying.

Change back to a mixed number.

Try some for yourself

13. (a) $1\frac{1}{3} \times 2\frac{1}{2}$ (d) $12\frac{1}{2} \div 2\frac{3}{4}$

(b) $3\frac{1}{6} \times 1\frac{1}{5}$ (e) $1\frac{1}{2} \times 3\frac{1}{6} \div 1\frac{1}{8}$

(c) $2\frac{1}{4} \div 1\frac{1}{3}$

Remember BO<u>D</u><u>M</u>AS!

Converting fractions to decimals

Can you explain to me the connection between fractions and decimals? I know there is some link but I'm not sure what it is. How can I change a fraction into a decimal?

Fractions and decimals are just different ways of writing the relationship between two numbers. Converting fractions to decimals is very easy when you use a calculator.

How are fractions and decimals connected?

Let's look at the fraction $\frac{5}{4}$.

This is a way of writing $5 \div 4$.

When we divide 5 by 4 we get the answer 1.25

$\boxed{5}\ \boxed{\div}\ \boxed{4}\ \boxed{=}\ \ 1.25$ (Your calculator works in decimals.)

1.25 is the decimal equivalent of the improper fraction $\frac{5}{4}$.

Every fraction has a decimal equivalent although some may be infinitely long or recurring.

$$\frac{22}{7} = 3.142\,857\,143 \ldots$$

$$\frac{1}{6} = 0.166\,666\,666 \ldots$$

This means that the decimal equivalents of some fractions are accurate only to a given number of decimal places or significant figures.

Converting fractions to decimals is easy when you use your calculator.

Convert $1\frac{3}{4}$ to a decimal.

$$1\frac{3}{4} = 1 + \frac{3}{4}$$

When using the calculator it is easier to work out the fraction first, then add the whole number.

$\boxed{3}\ \boxed{\div}\ \boxed{4}\ \boxed{+}\ \boxed{1}\ \boxed{=}\ \ 1.75$

What happens if you try to do it the other way?

$\boxed{1}\ \boxed{+}\ \boxed{3}\ \boxed{\div}\ \boxed{4}\ \boxed{=}$

A scientific calculator should work out the division first.
A simple calculator will work from left to right giving the answer 1.
($1 + 3 = 4, 4 \div 4 = 1$)

You are unlikely to be asked to change decimals to fractions. The quickest way to perform calculations is by using your calculator and your calculator works in decimals!

▽ Try some for yourself

14. Convert these fractions to decimals.

(a) $3\frac{1}{6}$ (b) $1\frac{15}{16}$ (c) $4\frac{1}{2}$ (d) $7\frac{3}{9}$

What have you learnt about fractions?

The most important thing I have learnt is not to be afraid of fractions. I've learnt lots of other things too.

- ✓ I can write one number as a fraction of another.
- ✓ I can identify equivalent fractions.
- ✓ I can add, subtract, multiply and divide fractions.
- ✓ I can add, subtract, multiply and divide mixed numbers.
- ✓ I can change fractions to decimals by using my calculator.

Further Questions

1. A mail-order firm receives 520 telephone calls in one day. A quarter of the calls are enquiries, five-eighths are orders and the remainder are from suppliers. How many calls were:
   (a) enquiries
   (b) orders
   (c) from suppliers?

2. An illness lasting 92 days required 23 days convalescing followed by 14 days holiday. What fraction of the total time was spent convalescing?

3. The Council Tax is split into three main areas of expenditure: three-tenths is spent on capital financing, one third on staff costs and the rest on services, supplies, buildings and vehicles. What fraction is spent on services, supplies, buildings and vehicles? If the total tax per household is £416.00, how much does each household contribute to each of the three main areas?

4. A restaurant is required to prepare 432 portions of turkey for a Christmas function. Twenty-four portions can be obtained from one bird.

   (a) What fraction of the total number of portions is supplied from one bird?

   (b) What fraction of the total number of portions is supplied from five birds?

5. The same size of ladies' shoes are marked:

   Size 6   British
   Size 7   American
   Size $39\frac{1}{2}$ Continental

A market stall has 46 pairs of these shoes marked 6, 10 pairs marked 7 and 73 pairs marked $39\frac{1}{2}$.

What fraction of the total number of shoes are

   (a) British size,

   (b) American size,

   (c) Continental size?

6. (a) A double-decker bus has seats for 32 passengers downstairs and 35 people upstairs. On a journey three-quarters of the seats downstairs and three-fifths of the seats upstairs are occupied. How many passengers are on the bus?

   (b) On the return journey there are 17 passengers upstairs and 25 downstairs. What fraction of the total number of seats are unoccupied?

7. Newspapers are delivered, on request, to guests at an hotel. One-fifth of the guests take paper A, one quarter take paper B, and three-eighths take paper C and one-tenth request paper D. What fraction of rooms do not receive a newspaper?

8. For the Spring Sale, a store reduces the prices of certain goods by one-third. The list shows the normal prices of goods. Calculate the sale prices.

Sweatshirts	£16.99
Grandad tops	£12.99
Shorts	£ 9.99
Jeans	£29.99
Sweaters	£39.99
Bodies	£21.99
Leggings	£12.99
Tunics	£14.99

9. A shop assistant is paid at an hourly rate of £3.72. If she works late one evening per week her rate increases to time and one half. When she works on a Bank Holiday she is paid double time. Calculate her pay based on an 8-hour working day for the week when the Monday was a Bank Holiday and she worked an extra three hours for late night opening. She worked for five days in total.

10. A batch of electrical resistors were checked and $\frac{1}{25}$ of the batch were found to be faulty. The total number of resistors in the batch was 500. Calculate:

    (a) the number of faulty resistors

    (b) the number of satisfactory resistors

    (c) the fraction of the batch that was satisfactory.

Solutions

1. (a) $\frac{3}{4}$ (b) $\frac{5}{16}$ (c) $\frac{2}{3}$

2. (a) $\frac{5}{16}$ (b) $\frac{3}{8}$ (c) $\frac{1}{2}$

3. (a) $\frac{4}{5}$ (b) $\frac{4}{10} = \frac{2}{5}$ (c) $\frac{4}{15}$ (d) $\frac{10}{25} = \frac{2}{5}$

    (b) and (d) are equivalent.

4. (a) $\frac{2}{7}, \frac{\cancel{4}^2}{\cancel{28}_{14}} = \frac{\cancel{2}^1}{\cancel{14}_7} = \frac{1}{7}$ (b) $\frac{3}{16}, \frac{6}{8} = \frac{12}{16}$

    (c) $\frac{\cancel{4}^2}{\cancel{6}_3} = \frac{2}{3}, \frac{2}{3}$ $\frac{4}{6}$ and $\frac{2}{3}$ are equivalent.

5. (a) $\frac{\cancel{15}^{\cancel{3}^1}}{\cancel{30}_{\cancel{6}_2}} = \frac{\cancel{3}^1}{\cancel{6}_2} = \frac{1}{2}$ (b) $\frac{15}{60} = \frac{3}{12} = \frac{1}{4}$

    (c) $\frac{25}{100} = \frac{5}{20} = \frac{1}{4}$ (d) $\frac{54}{72} = \frac{27}{36} = \frac{9}{12} = \frac{3}{4}$

6. (a) $\frac{3}{10} + \frac{5}{10} = \frac{8}{10} = \frac{4}{5}$

    (b) $\frac{1}{3} + \frac{1}{6} = \frac{2}{6} + \frac{1}{6} = \frac{3}{6} = \frac{1}{2}$

    (c) $\frac{8}{9} + \frac{1}{5} = \frac{40 + 9}{45} = \frac{49}{45}$

7. (a) $\frac{9}{10} - \frac{5}{10} = \frac{4}{10} = \frac{2}{5}$

    (b) $\frac{11}{32} - \frac{7}{32} = \frac{4}{32} = \frac{1}{8}$

    (c) $\frac{7}{16} - \frac{2}{16} = \frac{5}{16}$

    (d) $\frac{5}{8} - \frac{2}{7} = \frac{35 - 16}{56} = \frac{19}{56}$

    (e) $\frac{11}{12} - \frac{4}{9} = \frac{33 - 16}{36} = \frac{17}{36}$

    (f) $\frac{5}{16} + \frac{1}{3} - \frac{1}{6} = \frac{15 + 16 - 8}{48} = \frac{23}{48}$

8. (a) $\frac{11}{12} \times \frac{3}{11} = \frac{1}{4}$ (d) $12 \times \frac{3}{4} = 9$

    (b) $\frac{5}{6} \times \frac{9}{10} = \frac{3}{4}$ (e) $\frac{1}{2} \times \frac{4}{5} = \frac{2}{5}$

    (c) $\frac{17}{33} \times \frac{12}{34} \times \frac{3}{4} = \frac{3}{22}$ (f) $720 \times \frac{5}{8} = 450$ male

    $720 \times \frac{3}{8} = 270$ female

9. (a) $10 \div \frac{1}{2} = 10 \times \frac{2}{1} = 20$

    (b) $\frac{2}{5} \div \frac{5}{7} = \frac{2}{5} \times \frac{7}{5} = \frac{14}{25}$

    (c) $\frac{16}{9} \div \frac{4}{3} = \frac{\cancel{16}^4}{\cancel{9}_3} \times \frac{\cancel{3}^1}{\cancel{4}_1} = \frac{4}{3}$

    (d) $\frac{14}{8} \div \frac{7}{16} = \frac{\cancel{14}^2}{\cancel{8}_1} \times \frac{\cancel{16}^2}{\cancel{7}_1} = 4$

10. (a) $3\frac{5}{6} = \frac{23}{6}$
    (b) $6\frac{1}{4} = \frac{25}{4}$
    (c) $2\frac{28}{33} = \frac{94}{33}$
    (d) $1\frac{9}{16} = \frac{25}{16}$

11. (a) $\frac{23}{6} = 3\frac{5}{6}$
    (b) $\frac{25}{4} = 6\frac{1}{4}$
    (c) $\frac{84}{31} = 2\frac{22}{31}$
    (d) $\frac{15}{4} = 3\frac{3}{4}$

12. (a) $1\frac{5}{8} + 2\frac{1}{4} = 3\frac{7}{8}$
    (b) $5\frac{1}{6} + \frac{3}{8} = 5\frac{13}{24}$
    (c) $2\frac{1}{5} + 1\frac{3}{10} = 3\frac{1}{2}$
    (d) $5\frac{3}{4} - 4\frac{1}{8} = 1\frac{5}{8}$
    (e) $4\frac{1}{2} - 1\frac{3}{4} = 2\frac{3}{4}$
    (f) $2\frac{1}{3} - 1\frac{1}{6} = 1\frac{1}{6}$

13. (a) $1\frac{1}{3} \times 2\frac{1}{2} = 3\frac{1}{3}$
    (b) $3\frac{1}{6} \times 1\frac{1}{5} = 3\frac{4}{5}$
    (c) $2\frac{1}{4} \div 1\frac{1}{3} = 1\frac{11}{16}$
    (d) $12\frac{1}{2} \div 2\frac{3}{4} = 4\frac{6}{11}$
    (e) $1\frac{1}{2} \times 3\frac{1}{6} \div 1\frac{1}{8} = 4\frac{2}{9}$

14. (a) $3\frac{1}{6}$ = 3.16667 to 5 d.p.
       = 3.17 to 2 d.p.
    (b) $1\frac{15}{16}$ = 1.9375
    (c) $4\frac{1}{2}$ = 4.5
    (d) $7\frac{3}{9}$ = 7.3333 . . .

4 Decimals

> Decimals are quite easy. After all, we use them all the time for money, lengths and other measures. I'm fairly confident about working with them as long as I have my calculator!

> It's quite true that the metric system and decimal currency mean that we all use decimals all the time. I think it might be useful to do a quick bit of revision of decimals – especially looking at doing decimal arithmetic without a calculator.

What is a decimal?

When you divide a number you are often left with a remainder. When you use your calculator for division this remainder appears in decimal form.

$$18 \div 7 = 2.571\,428\,571$$

Decimal part of remainder

$$18 \div 7 = 2\tfrac{4}{7}$$

Fractional remainder

$$18 \div 7 = 2 \quad \text{Remainder } 4$$

These are *all* correct answers. We are only interested in the decimal answer.

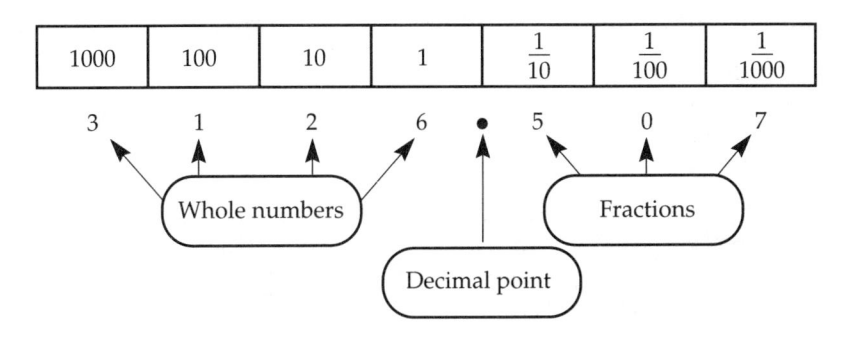

| 1000 | 100 | 10 | 1 | $\frac{1}{10}$ | $\frac{1}{100}$ | $\frac{1}{1000}$ |

3 1 2 6 . 5 0 7

Whole numbers — 1, 2, 6

Decimal point — .

Fractions — 5, 0, 7

53

The decimal point is used to separate whole numbers from fractions.
$$3126.507 = 3126 + \frac{5}{10} + \frac{0}{100} + \frac{7}{1000}$$

The position of the digits in relation to the decimal point gives us the **place value**.

$$0.5 = \frac{5}{10}$$
$$0.50 = \frac{5}{10} + \frac{0}{100}$$
$$0.507 = \frac{5}{10} + \frac{0}{100} + \frac{7}{1000}$$

The zero between 5 and 7 is very important. It tells us that there are no hundredths. We cannot leave it out or the number would change. We should then have $\frac{5}{10} + \frac{7}{100}$ which is not the same.

Place value is important – zeros have meaning and must not be omitted.

Place value is very useful when we wish to compare decimal numbers.

Is 2.9 greater than 2.897?

1	$\frac{1}{10}$	$\frac{1}{100}$	$\frac{1}{1000}$
2 .	9	0	0
2 .	8	9	7

900 is greater than 897.

So 2.9 is greater than 2.897

To use mathematical language, we would write

$$2.9 > 2.897$$

\> means greater than

We could write it another way. We could say

2.897 is less than 2.9

$$2.897 < 2.9$$

< means less than

These symbols are a form of mathematical shorthand. They are often used in mathematical texts and you should be aware of their meanings.

> means greater than
< means less than

Metric measurements are based on multiples of 10. Prefixes such as kilo-, centi-, milli-, etc. have definite meanings.

kilo- means 1000
deci- means one tenth or $\frac{1}{10}$
centi- means one hundredth or $\frac{1}{100}$
milli- means one thousandth or $\frac{1}{1000}$

1 kilogram (kg) = 1000 grams (g)
1 kilometre (km) = 1000 metres (m)
1 metre (m) = 100 centimetres (cm)
            = 1000 millimetres (mm)

We can convert from one unit to another by multiplying or dividing by 10, 100, or 1000, etc.

Remember: When you multiply a decimal by 10, it's the same as moving the decimal point one place to the right.

$$2.34 \times 10 = 23.4$$

When you multiply by 100, you move the decimal point *two* places to the right.

$$2.34 \times 100 = 234.$$

When you *divide* by 10, you move the decimal point one place to the *left*.

$$2.34 \div 10 = 0.234$$

When you divide by 100, you move the decimal point *two* places to the left.

$$2.34 \div 100 = 0.0234$$

Working with decimals

Adding decimals

I think you rely on your calculator too much. It would be useful to be able to do simple decimal arithmetic without your calculator. I'm sure you can add decimals without using a calculator.

I can add decimals if I really need to. I can add up the cost of items of shopping to make sure I'm not charged too much. I'm not very good at setting out sums on paper.

When we add decimals it is best to write the numbers in a column with the **decimal points** in line.

$$5.54 + 10.32 + 6.04$$

```
   5.54
  10.32
   6.04
  ₁  ₁
  21.90
```

Lining up the decimal points makes the sum look neat and avoids confusion about place value.

Subtracting decimals

The same rule applies to the subtraction of decimals as it does to the addition of decimals.

> Decimal points should be in line.

Subtract 94.56 from 163.87

$$\begin{array}{r} 163.87 \\ -94.56 \\ \hline 69.31 \end{array}$$

If you want to check that your answer is correct just add the answer to the number being subtracted. The result should equal the number from which the subtraction was made.

$$\begin{array}{r} 94.56 \\ +69.31 \\ \hline 163.87 \end{array}$$

This is a good habit to acquire.

Try some for yourself
1. (a) $10.38 + 16.007 + 3.1 + 30$
   (b) $1.634 + 10.01 + 33.333$
   (c) $1096.4 - 98.32$
   (d) $21 - 13.02$

Dividing decimals

> I really do not understand how to divide one decimal by another decimal. I just can't do it without my calculator.

> Well, this may seem a strange thing to say, but the easiest way to divide by a decimal is not to divide by a decimal!

> That doesn't make sense at all! How can you not divide by a decimal when you are dividing by a decimal?

> I see you doubt my logic! What I mean is that to avoid dividing by a decimal we change the problem. We can divide by a whole number so that's what we do. We change the decimal into a whole number.

This is how we do it.

If we take two numbers,   6 and 3   6 ÷ 3 = 2
Multiply both numbers by 10,   60 and 30   60 ÷ 30 = 2
Multiply both numbers by 100, 600 and 300   600 ÷ 300 = 2

As long as both numbers are multiplied by the same number the result will always be the same.

We apply this system to decimals to get rid of those tiresome decimal points.

$$14.84 \div 0.4$$

Multiply both numbers by 10.

$$148.4 \div 4$$

We can now divide 148.4 by 4 because we are dividing by a whole number.

$$148.4 \div 4 = 37.1$$

Now let's try $0.538 \div 0.02$

Multiply both numbers by 10.

$$5.38 \div 0.2$$

We are still dividing by a decimal.

Multiply both numbers by 10 again.

$$53.8 \div 2$$

Now we are dividing by a whole number.

$$53.8 \div 2 = 26.9$$

$$2 \overline{)53.8}$$
$$26.9$$

> So the idea is to multiply both numbers by 10 as many times as it takes to make sure we are dividing by a whole number?

> That's right. It doesn't matter if the number being divided is a decimal – that's no problem. Dividing by a decimal is a problem. That is what we want to avoid.

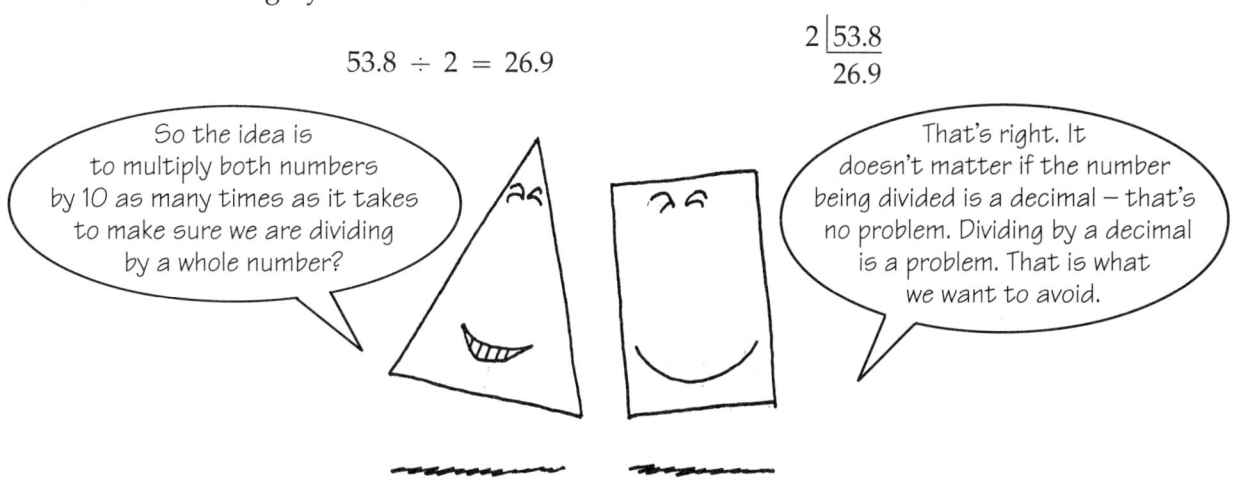

Try some for yourself

2. (a) $7.56 \div 9$ (e) $0.42 \div 0.07$
(b) $19.75 \div 5$ (f) $1.69 \div 0.13$
(c) $6.44 \div 4$ (g) $0.04 \div 0.1$
(d) $123.65 \div 5$ (h) $25.03 \div 0.005$

Multiplying decimals

> I feel reasonably happy about adding and subtracting decimals. Multiplication is rather hazy. Isn't there a funny rule about ignoring the decimal point then putting it somewhere in the answer?

> You have obviously remembered only part of what you had learned about multiplying decimals. You have forgotten the rule that tells you where to put the decimal point after the multiplication.

Let's look at a few examples to see if we can find a pattern.

$0.2 \times 2 = 0.4$
$0.2 \times 0.2 = 0.04$
$0.2 \times 0.02 = 0.004$
$0.2 \times 0.002 = 0.0004$
$0.2 \times 0.0002 = \quad ?$

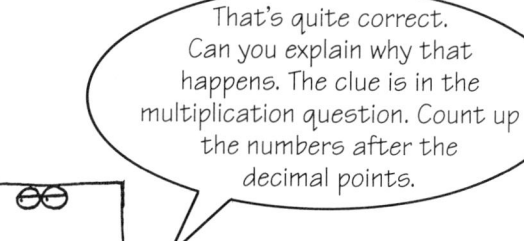

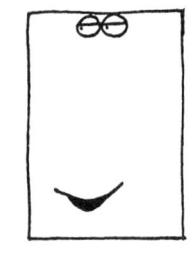

> The answer must be 0.00004. Every time you added a 0 in the decimal you were multiplying by, the decimal point moved 1 place to the left in the answer.

> That's quite correct. Can you explain why that happens. The clue is in the multiplication question. Count up the numbers after the decimal points.

> Thanks for the clue. I think I begin to see a pattern. The number of digits after the decimal point in the answer is the same as the total number of digits after the decimal point in the multiplication.

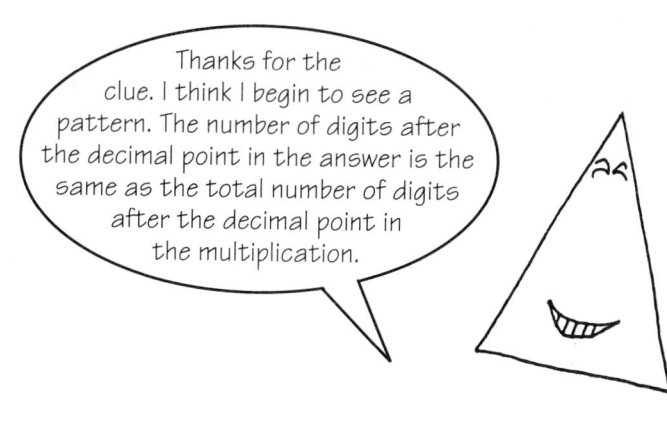

0.2 × 2	1 digit	0.2
0.2 × 0.2	2 digits	0.2 and 0.2
0.2 × 0.02	3 digits	0.2 and 0.02
0.2 × 0.002	4 digits	0.2 and 0.002
0.2 × 0.0002	5 digits	0.2 and 0.0002

That's exactly the answer. To summarise the rule:

> The total number of decimal places (digits to the **right** of the decimal point) in the question is equal to the number of decimal places in the answer.

Accuracy — decimal places

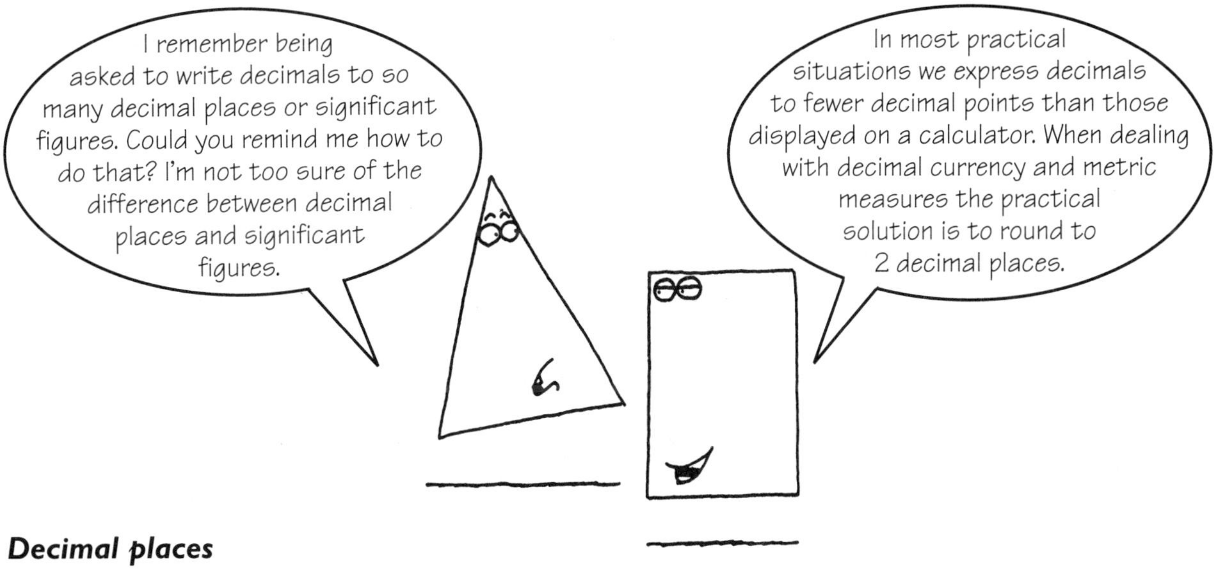

> I remember being asked to write decimals to so many decimal places or significant figures. Could you remind me how to do that? I'm not too sure of the difference between decimal places and significant figures.

> In most practical situations we express decimals to fewer decimal points than those displayed on a calculator. When dealing with decimal currency and metric measures the practical solution is to round to 2 decimal places.

Decimal places

£200 ÷ 9 = £22.22 (2222) These have no meaning in practical terms.

= £22 + 22 pence

= £22.22 to 2 decimal places

'decimal places' is usually written as d.p.

Round   2.571 428 571   to 1, 2 and 3 decimal places.

The rules for rounding decimals are the same as those for rounding whole numbers.

> 0, 1, 2, 3, 4 rounds down

> 5, 6, 7, 8, 9 rounds up

2.571 428 571   to 1 d.p.

> Look at the second decimal digit. 7 tells us to round up.

= 2.6 to 1 d.p.

2.571 428 571   to 2 d.p.

> The third decimal digit is 1 so round down.

= 2.57 to 2 d.p.

2.571 428 571   to 3 d.p.

> The fourth decimal digit is 4 so round down.

= 2.571 to 3 d.p.

Round   2.298   to 2 d.p.

> The third decimal digit is 8 so round up.

=   2.30   to 2 d.p.

> When 9 is rounded up to 10, the first decimal digit is rounded up.

Accuracy – significant figures

I think I remember decimal places quite well and that reminder has jogged my memory. I always had problems with significant figures. I didn't understand what they were.

*Decimal places are obvious – places following the decimal point. Significant figures include the digits **before** the decimal point. The important point is that the figures must be **significant.***

That does not help me much. What makes a figure significant? What figures are insignificant? I think I shall need more help with these.

*All figures are significant apart from **'leading zeros'**. That means that if a zero (0) comes before any other digit it is **not** significant. It doesn't matter if it's before or after the decimal point. A zero (0) is only significant when it follows another digit.*

Round (0.00)34   to 1 significant figure

> Leading zeros are insignificant figures.

0.0034 = 0.003 to 1 significant figure.
3 is the first **significant** figure.

Round 1 0 9 . 3 4   to 4 significant figures
    1st 2nd 3rd 4th   = 109.3 to 4 sig. figs./s.f.

all significant

'significant figures' is usually written s.f. or sig.figs.

> The rules for rounding significant figures are the same as those for decimal places and whole numbers.

£9124 = £9000 to 1 sig.fig.

Try some for yourself

3. Round each of these numbers to (i) 1 d.p. (ii) 2 d.p.

(a) 234.321 m   (d) 0.654 m
(b) £200.059   (e) £0.089
(c) 2.267 kg   (f) 1.825 kg

4. Round each of these numbers to (i) 1 s.f. (ii) 2 s.f. (iii) 3 s.f.

(a) 0.06752   (c) 89 165.0
(b) 91.82   (d) 0.000 1903

More calculator information

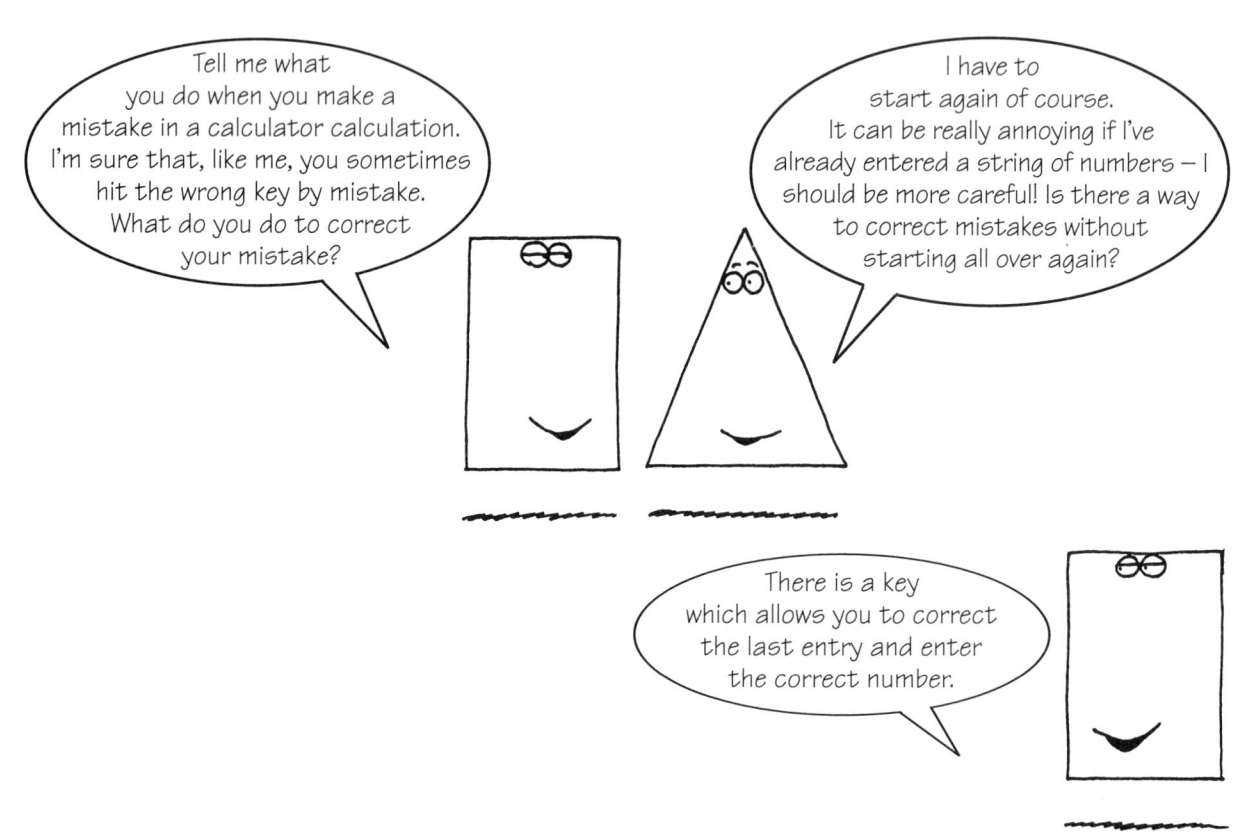

Tell me what you do when you make a mistake in a calculator calculation. I'm sure that, like me, you sometimes hit the wrong key by mistake. What do you do to correct your mistake?

I have to start again of course. It can be really annoying if I've already entered a string of numbers – I should be more careful! Is there a way to correct mistakes without starting all over again?

There is a key which allows you to correct the last entry and enter the correct number.

|CE| or |C| Clear entry. This key allows you to clear an incorrectly
Sharp Casio entered number and enter the correct number. The common
labellings are CE or C but it's always sensible to check your
maker's handbook in case your calculator is different.

Try using the appropriate key on your calculator.

Enter 21 × 24 then correct 24 to 25.

Key sequence: |2| |1| |×| |2| |4| |CE| |2| |5| |=|

Note: there is no need to press |×| key again.

This error correction facility is very useful, particularly if you make a mistake when entering a long string of numbers.

▽ Try some for yourself

Use your calculator to work out:

5. (a) 23 + 25 + 27 + 28   Correct 28 to 30.

(b) 3 × 4 × 5 × 7   Correct 7 to 6.

(c) 234 − 227   Correct 227 to 272.

(d) 652 ÷ 136   Correct 136 to 163.

That's really clever. I shall make good use of that key. I wish I had known about it when I started using a calculator. What other useful keys should I know about?

I don't expect you know how to use the calculator's memory. Most students have no idea how powerful their calculators are. The memory is incredibly useful for all sorts of calculations.

Using the memory

Your calculator may have one or more memories. The memory can be used to store parts of a calculation for use at a later stage.

Once again, check in *your* maker's handbook to see what your calculator's memory keys look like, and how they are used, if your calculator does not have the following keys.

|Min| or |x→m| Stores the number being displayed.

|MR| or |RM| Recalls the memory, displaying its contents.

|M+| Adds the number being displayed to the memory.

|M−| Subtracts the number being displayed from the memory.

Whatever memory keys your calculator has, you will also need to find out how to clear the memory. Make sure you know how to do this. Check that you can by carrying out the following steps.

1. Store the number 6 in the memory.

$\boxed{6}$ $\boxed{\text{Min}}$ The display should indicate 6.

2. Clear the display by using clear entry key $\boxed{\text{CE}}$ or $\boxed{\text{C}}$.

3. Recall the memory to make sure there is something in it.

$\boxed{\text{MR}}$ recalls the memory and the display should indicate 6.

4. Clear the memory.

$\boxed{0}$ $\boxed{\text{x} \to \text{m}}$ or $\boxed{0}$ $\boxed{\text{Min}}$ enters zero in the memory.

5. Check that the memory is clear.

$\boxed{\text{MR}}$ or $\boxed{\text{RM}}$ recalls the memory and the display should indicate 0.

It is always sensible to clear both the display and the memory before starting any new calculation. Many calculators can now continue to store a number in the memory after the calculator has been switched off.

Using the memory in calculations

The memory can often be used when a calculation involves brackets. The brackets indicate which part you should work out first. Instead of using the calculator's brackets, you can work out that part first and store it in the memory, recalling it again when you need it.

Example

Work out $21 + (192 - 57)$.

Solution

1. Work out the term in brackets first, storing the result in the memory.

$\boxed{1}$ $\boxed{9}$ $\boxed{2}$ $\boxed{-}$ $\boxed{5}$ $\boxed{7}$ $\boxed{=}$ $\boxed{\text{Min}}$

2. Add 21 to the memory.

$\boxed{2}$ $\boxed{1}$ $\boxed{\text{M+}}$

3. Recall the answer.

$\boxed{\text{RM}}$ 156

The entire calculation can be performed in one key sequence.

$\boxed{1}$ $\boxed{9}$ $\boxed{2}$ $\boxed{-}$ $\boxed{5}$ $\boxed{7}$ $\boxed{=}$ $\boxed{\text{Min}}$ $\boxed{2}$ $\boxed{1}$ $\boxed{\text{M+}}$ $\boxed{\text{RM}}$

▽ Try some for yourself

6. Use the memory of your calculator to work out each of these.

(a) (i) $6 + (5 - 3)$

(ii) $9 + (3 \times 5)$

(iii) $42 - (6 \times 4)$

(iv) $96 \div (24 - 16)$

(b) (i) 6.321 + (17.6 − 24.3)

(ii) 14.7 ÷ (0.2 + 10.62)

(iii) 0.12 + (0.001 × 14.6)

(iv) 12.01 × (16.5 − 17.4)

The memory can also be useful when you have problems involving fractions.

$\frac{3}{4}$ is another way of writing 3 ÷ 4

Work out $\frac{48}{2 + 6}$

That is another way of writing 48 ÷ (2 + 6).

Work out (2 + 6) then divide the answer into 48.

| 2 | + | 6 | Min | 4 | 8 | ÷ | MR | = |

Notice that we stored the answer to 2 + 6 in the memory. We then entered 48 followed by the ÷ sign and then recalled the memory. We were telling the calculator to divide 48 by the number in the memory.

Try some for yourself

7. Use the memory to help you to work out each of the following.

(a) $\frac{18}{4 + 2}$

(b) $\frac{153.6}{22.1 + 13.7}$

(c) $\frac{85 - 27}{13 + 16}$

(d) $\frac{2.3 \times 6.5}{3.7 \times 8.1}$

(e) $\frac{55}{3.2 \times 9.7}$

That seems a rather long-winded way to do fractions. It's easier to work out the numerator and denominator and then divide. Surely my calculator has an easier way of doing this?

Most calculators and all scientific calculators should have a reciprocal key. Do you remember what we discovered about reciprocals when we were revising fractions?

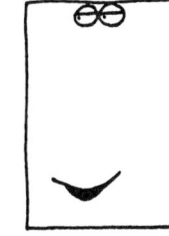

The reciprocal key

We used reciprocals when we were dividing by fractions. When we turned the fraction upside down we got the reciprocal. I also remember that a fraction times its reciprocal always equals 1.

I'm impressed by your memory and your understanding. The reciprocal key on the calculator gives you the reciprocal of the number in the display. This can be very useful when you are performing fraction calculations. I'll explain.

| 1/x | The reciprocal key.

Let's find the reciprocal of $\frac{1}{2}$.

| 1 | ÷ | 2 | = | 1/x | 2

Try some for yourself

8. Try out your reciprocal key on these numbers.

(a) 5

(b) 100

(c) − 10

(d) 0·125

You can use the reciprocal key to work out problems like

$$\frac{25}{2+3} \quad \text{because} \quad \frac{25}{2+3} = 25 \times \frac{1}{(2+3)}$$

$\frac{1}{(2+3)}$ is the reciprocal of $(2+3)$ because $\frac{1}{5} \times 5 = 1$

Try entering this sequence in your calculator.

| 2 | + | 3 | = | 1/x | × | 25 | 5

We must work out the denominator (5) first then we take the reciprocal to give us the fraction ($\frac{1}{5}$). That is then multiplied by the numerator (25).

▽ Try some for yourself

9. (a) $\dfrac{18}{4+2}$ (d) $\dfrac{2.3 \times 6.5}{3.7 \times 8.1}$

(b) $\dfrac{153.6}{22.1 + 13.7}$ (e) $\dfrac{55}{3.2 \times 9.7}$

(c) $\dfrac{85 - 27}{13 + 16}$

To work out calculations involving division you can use either the memory or the reciprocal key. It doesn't matter which method you use as long as you are happy with it. Choose the one you feel more confident with.

What have you learnt about decimals?

I was pretty confident about working with decimals when I started. I have certainly learnt a lot more about handling them.

- [✓] I understand what place value is.
- [✓] I can multiply or divide a decimal number by multiples of 10 without using my calculator.
- [✓] I can add, subtract, multiply and divide decimal numbers without using my calculator.
- [✓] I can round a decimal number to a given number of decimal places.
- [✓] I can round a decimal number to a given number of significant figures.
- [✓] I can use the memory on my calculator.
- [✓] I understand how to use the reciprocal key on my calculator.

Further Questions

1. Yang and Paul decide to order a Chinese take-away. Calculate the total cost of the food they ordered.

1 chicken and noodle soup	£0.95
1 chicken and sweetcorn soup	£1.05
1 portion egg fried rice	£1.15
1 portion Chow Mein	£1.65
Chicken with cashew nuts	£3.25
Shrimp with mushrooms	£2.85
Prawn crackers	£0.90

2. The Wong family are going to Euro Disney in France. The party is made up of the parents and two sons. The prices for seven nights accommodation are given as follows:

Each adult	£498.00
Each child	£298.80

   Calculate the total cost of the package.

3. Driving lessons for students are offered at a special rate.

   £11.50 for one lesson per week.
   £ 6.50 for the first lesson.
   £ 6.50 for the fifth lesson.

   Find the cost of a series of eight lessons.

4. Lloyd wants to buy a personal computer. He is quoted prices for the different items of hardware he requires. He can afford £1000 but needs to know the total he will have to pay for the system he wants.

386SX PC	£703.83
Printer	£233.83
Box of diskettes	£ 12.99

   How much money will he have after he has paid for the hardware?

5. Judith has moved into a new house with a new garden. She wants to plant the garden with bulbs and perennials. She also wants to plant her window boxes. Find the total cost of all her purchases from the garden centre.

30 mixed perennials	£12.99
12 lavender bushes	£ 8.99
500 bulbs (assorted)	£14.99
36 begonias	£ 8.99
36 busy lizzies	£ 9.99

6. David has saved his money from his part-time job to buy a CD player and some CDs. How much change will he have from his £250 savings?

Portable CD player	£149.99
Mains adaptor	£ 21.99
6 CDs at	£ 13.49 each

7. Irina and her friend Sarah decide to have an Indian take-away. They can afford £12 between them. Will this be enough to buy the food? They want 3 papadums at 30p each, Tandoori Chicken at £3.95, Chicken Dhansak at £3.85, Pilau Rice at £1.00, one Nan bread costing 95p and 1 Chapati which costs 70p.

8. The table shows the number of laying hens on six farms.

Hampton	2138
Blue Ridge	1654
Three Oaks	1903
Orange	1551
Pine tree	2004
Wingland	1768

   Rewrite the table giving the number of hens correct to two significant figures.

9. An ambulance answering an emergency call took 6.4 min to travel the distance of 4.08 miles between the ambulance depot and the patient's home. The time taken to get the patient into the ambulance was 3.89 min and the journey of 6.37 miles from the patient's home to the hospital took 11.73 min.

   (a) For what time was the ambulance being driven?

   (b) How many miles did the ambulance cover?

   (c) What was the total time taken between the ambulance leaving its depot and arriving at the hospital?

10. The batting average of a cricketer is found from the formula

    $$\text{average runs} = \frac{\text{total runs}}{\text{number of finished innings}}$$

    Use this formula to find the batting averages, correct to 1 decimal place of the following players in a youth team.

Player	Total runs	Number of finished innings	Batting average
S. Gilbert	126	5	
T. Leigh	72	4	
A. Thompson	84	5	
C. Sainty	98	5	
D. Evans	22	3	
R. Fordham	36	3	

11. The table shows the distance in miles from London to five coach destinations. Rewrite the table giving the distances in kilometrees correct to 1 decimal place.

    1 mile = 1.61 kilometres

	Cambridge	Penzance	Newcastle	Aberdeen	Holyhead
London	54	281	273	492	359

12. Give the following figures correct to:

    (i) 2 decimal places   (ii) 2 significant figures.

    (a) 2.7938   (d) 0.045 78

    (b) 0.051   (e) 126.0918

    (c) 8.167

13. A businessman is taking a trip to the U.S.A. He changes £500 to dollars when the exchange rate is £1 = 1.45 dollars. He changes the dollars he has left on his return back to pounds at the rate of £1 = 1.49 dollars.

    (a) How many dollars did he receive for his £500?

    (b) He spent $397 on his trip. How many dollars remained?

    (c) How much did he receive when he changed the dollars back to pounds?

14. Use the relationship

    $$\text{current (amperes)} = \frac{\text{potential different (volts)}}{\text{resistance (ohms)}}$$

    to complete the following table:

Volts	Amperes	Ohms
12		10
24		7.5
6		12
240		50
18		8
7.5		25
110		40
36		30

Solutions

1. (a) 59.487 (b) 44.977 (c) 998.08 (d) 7.98
2. (a) 0.84 (e) 6
   (b) 3.95 (f) 13
   (c) 1.61 (g) 0.4
   (d) 24.73 (h) 5006
3. (a) (i) 234.3 m (ii) 234.32 m
   (b) (i) £200.1 (ii) £200.06
   (c) (i) 2.3 kg (ii) 2.67 kg
   (d) (i) 0.7 m (ii) 0.65 m
   (e) (i) £0.1 (ii) £0.09
   (f) (i) 1.8 kg (ii) 1.83 kg
4. (a) (i) 0.07 (ii) 0.068 (iii) 0.0675
   (b) (i) 90 (ii) 92 (iii) 91.8
   (c) (i) 90 000 (ii) 89 000 (iii) 89 200
   (d) (i) 0.0002 (ii) 0.000 19 (iii) 0.00 190
5. (a) 105 (b) 360 (c) -38 (d) 4
6. (a) (i) 8 (ii) 24 (iii) 18 (iv) 12
   (b) (i) -0.379 (ii) 1.359 (iii) 0.1346 (iv) -10.809
7. (a) 3 (d) 0.499
   (b) 4.29 (e) 1.772
   (c) 2
8. (a) 0.2 (b) 0.01 (c) -0.1 (d) 8
9. (a) 3 (d) 0.499
   (b) 4.29 (e) 1.772
   (c) 2

5 Ratios, Scales and Percentages

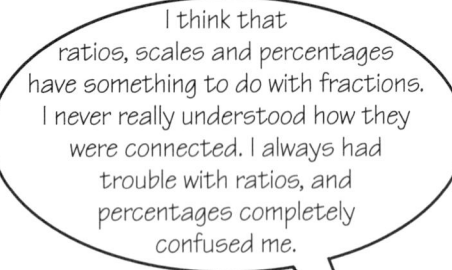

> I think that ratios, scales and percentages have something to do with fractions. I never really understood how they were connected. I always had trouble with ratios, and percentages completely confused me.

> Ratios are very like fractions. They give exactly the same information but they are written without a denominator. You have to work that out for yourself. It's quite easy when you know what to do.

What is a ratio?

A cocktail is a drink made up of different ingredients in a given ratio.

James Bond is famous for his fondness for a vodka martini, shaken, not stirred.

A vodka martini is made up of three parts vermouth to one part vodka.

The ratio of vermouth to vodka is $3:1$.

If we add the three parts of vermouth to the one part of vodka

$3 + 1 = 4$

We find we have 4 parts in total. We have the denominator for our fraction.

$\frac{3}{4}$ = vermouth

$\frac{1}{4}$ = vodka

Add together to make a vodka martini.

Square: Can you see the connection between the ratio 3 : 1 and the fractions $\frac{3}{4}$ and $\frac{1}{4}$? Can you see how ratios gave us the information to find the denominator?

Triangle: Well, when we added the two ratios together we found the denominator. 3 + 1 = 4. Is that all it is? Adding the ratios gives the denominator. Its so simple!

There are plenty of practical examples of ratios.

Shortcrust pastry is made in the ratio of 1 part fat to 2 parts flour.

$$1 : 2 \qquad 1 + 2 = 3$$
$$\tfrac{1}{3} \text{ fat} \qquad \tfrac{2}{3} \text{ flour}$$

If we want to make 12 lb of pastry we use the ratios to find how much fat and flour we need.

We need 2 parts of flour = $\frac{2}{3}$

We need 1 part of fat = $\frac{1}{3}$

We need $\frac{2}{3}$ of 12 lb of flour = $\cancel{12}^{4} \times \frac{2}{\cancel{3}_1}$ lb = 8 lb

We need $\frac{1}{3}$ of 12 lb of fat = $\cancel{12}^{4} \times \frac{1}{\cancel{3}_1}$ lb = 4 lb

The composition of bronze is in the ratio of 91 parts copper to 9 parts tin.

$$91 : 9 \qquad 91 + 9 = 100$$
$$\tfrac{91}{100} \text{ copper} \qquad \tfrac{9}{100} \text{ tin}$$

How much copper and tin will be needed to make 1000 tonnes of bronze?

We need 91 parts of copper = $\frac{91}{100}$

We need 9 parts of tin = $\frac{9}{100}$

$\cancel{1000} \times \dfrac{91}{\cancel{100}} = 910$ tonnes of copper

$\cancel{1000} \times \dfrac{9}{\cancel{100}} = 90$ tonnes of tin

Try some yourself

1. (a) Divide £1500 in the ratio $5:3:2$.

(b) The ratio of male to female students on a GNVQ course is $8:2$. If the total number of students is 150, how many are male and how many female?

Scales

Now I know how easy ratios are, I'd better revise scales. I know we use them for maps and plans and scale models. Are there any other uses? Are they easy to understand?

As you say, we use scales for maps, plans and models. We also use scaling to increase or decrease the amounts of ingredients in a recipe or an alloy. Let's look at scale drawings first.

Scale drawings

If I want to draw a plan of a room measuring 10 m by 5 m I cannot do it to size. I should need a piece of paper the size of the room. I have to draw the plan to scale. I need to decide on a suitable scale, so that the drawing will fit on the paper and be a reasonable size. My scale must relate 1 unit of my drawing to 1 unit of the actual measure. For the room plan, I could use a scale of 1 cm to represent 1 metre.

1 cm : 1 metre

It's better to have the two parts of the scale in the same units, so I change 1 m to 100 cm.

Scale 1 : 100

1 cm : 100 cm

Now I can omit the cms.

The scale is 1 : 100.

This tells me the relationship between the measurements on the plan and the measurements of the room.

When I measure the plan I find that the room is 10 cm × 5 cm. My scale tells me that the actual measurements are (10 × 100 cm) × (5 × 100 cm)
$$1000 \text{ cm} \times 500 \text{ cm}$$
$$10 \text{ m} \times 5 \text{ m}$$

▽ Try some for yourself

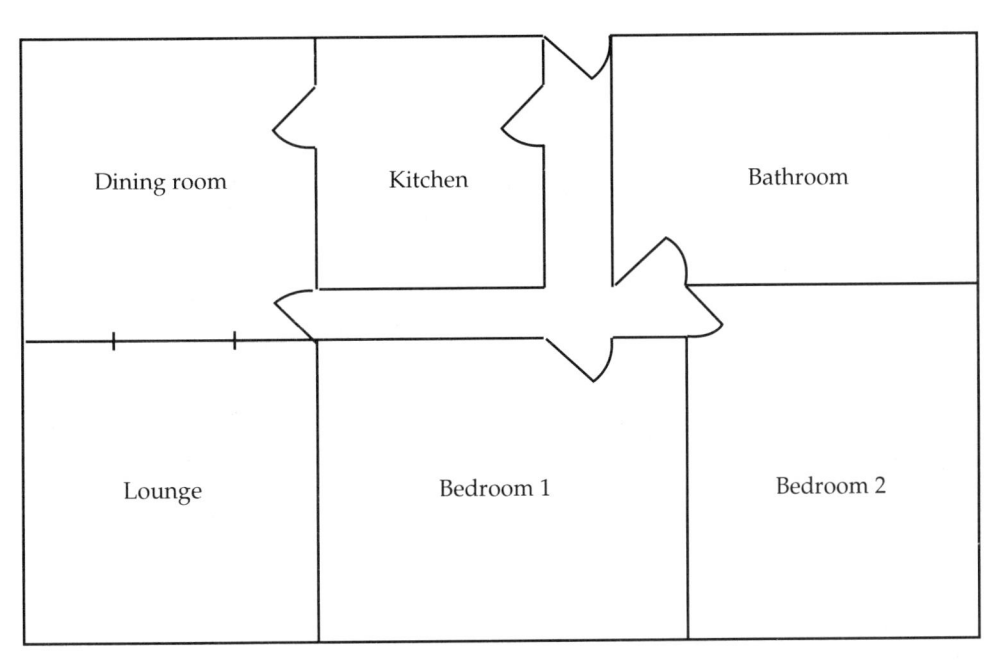

Scale 1 : 100

2. (a) What is the scale length of the house?

(b) What is the real length of the house?

3. (a) What is the scale width of the house?

(b) What is the real width of the house?

4. (a) What are the length and width of bedroom 1?

(b) What are the length and width of bedroom 2?

(c) Which is the larger room?

Map scales

Map scales work in exactly the same way as scales for plans. The difference is that the actual distance are much greater, so the scales are very large.

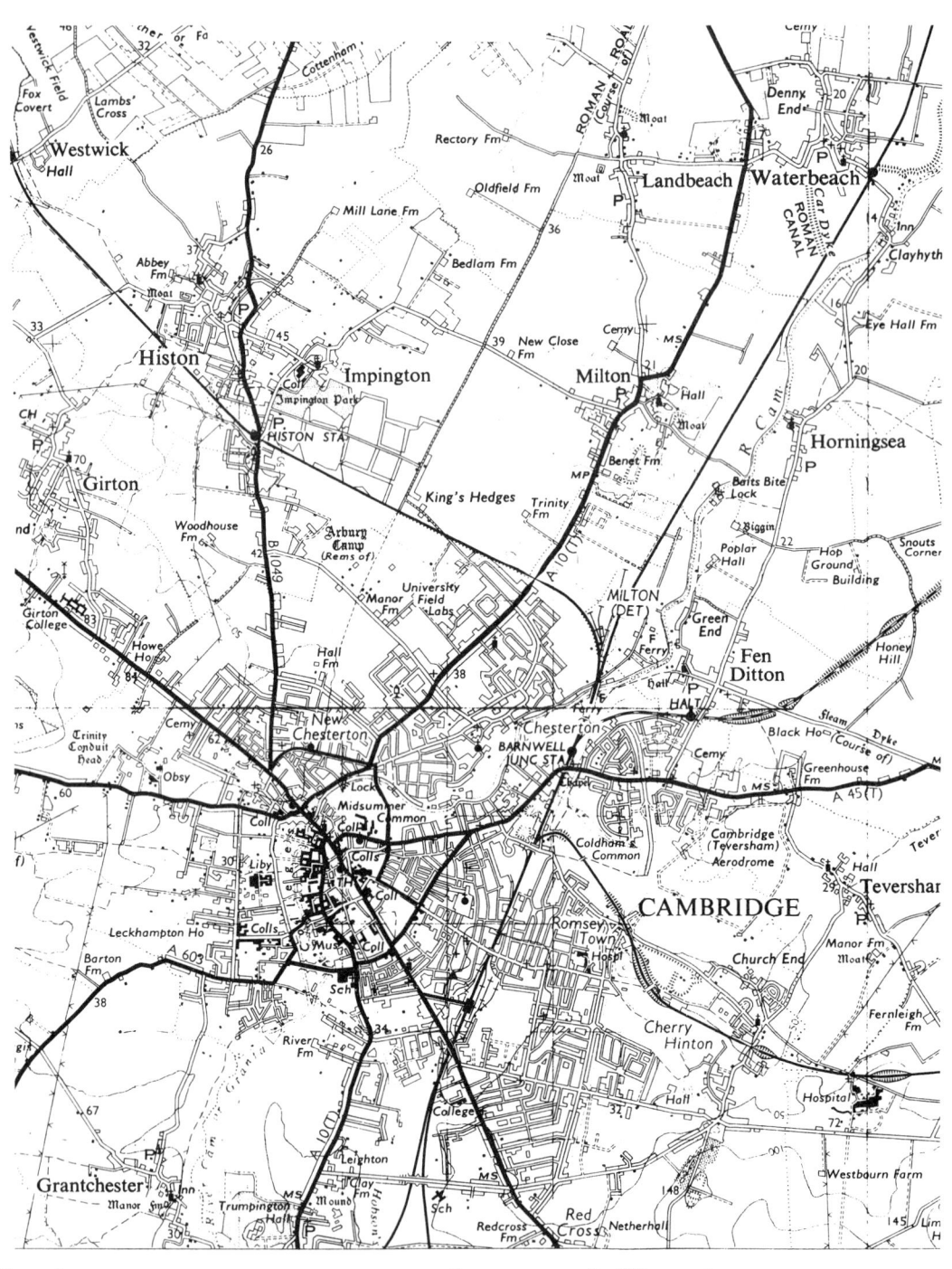

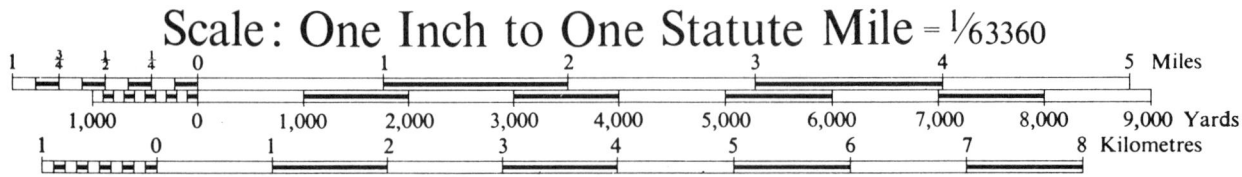

Scale: One Inch to One Statute Mile = 1/63360

76

This Ordnance Survey map has a scale of

        1 inch to 1 mile.
        1 inch to 63360 inches
        1 : 63360

In metric measure the scale is

        1.5 cm : 1 kilometre
        1.5 cm : 100 000 cm
        1.5   : 100 000

For every 1.5 cm on the map the distance on the ground is 1000 m or 100 000 cm.

The map shows Cambridge and the surrounding villages

Remember the scale is 1.5 : 100 000.

We are going to find the actual distance from Girton College to the junction with the B1049. (The points are each marked with a cross.)

On the map the distance is approximately 4 cm.

The actual distance is approximately 2667 metres = 2.7 km.

▽ Try some for yourself

   5. Measure the distance from Waterbeach railway station to Barnwell Junction Station. What is the actual distance?

Scaling quantities

Recipes for food or chemicals or metals involve the use of ingredients in certain quantities. It is often necessary to increase or decrease these quantities to satisfy a given need. This adjustment is carried out by scaling the quantities.

When making shortbread I find that:

        255 grams of plain flour
          85 grams of castor sugar
        170 grams of butter

will produce 510 grams of shortbread.

My student daughter wants three times that amount to feed her college friends. What quantities shall I use?

I want 3 × 510 grams, so I need to scale each quantity by 3.

        3 × 255 = 765 grams flour
        3 × 85 = 255 grams sugar
        3 × 170 = 510 grams butter

Hopefully, this will feed a few starving students for several weeks.

Some recipes give you the number of servings. If you are feeding more people (or fewer) than the suggested servings, you will need to scale your recipe.

Tomato and Leek Salad – Serves 4

> 450 g of tomatoes
> 100 g of leeks
> 60 millilitres dressing

What amounts of each ingredient will I need to serve 6 people?

To find the scale necessary I divide 6 by 4.

$6 \div 4 = 1.5$ I must scale by 1.5

$450 \times 1.5 \text{ g} = 675 \text{ g}$ or tomatoes
$100 \times 1.5 \text{ g} = 150 \text{ g}$ of leeks
$60 \times 1.5 \text{ ml} = 90 \text{ ml}$ of dressing

▽ Try some for yourself

6. When making bread 20 g of fresh yeast is added to 700 g of flour. What quantities of yeast and flour will be needed to make 1080 g of bread?

7. To make 10 g of a solder we use 6 g of lead and 4 g of tin.

What quantities of lead and tin will be needed to make 25 g of solder?

Percentages

I really do have terrible problems with percentages. I just didn't understand them at school. Now I find I need them for such a variety of problems. I really must come to grips with them.

Percentages are really not so complicated. They can be handled either as fractions or decimals. Now you can cope with fractions and decimals, percentages should be easier.

Percentages crop up in various aspects of everyday life. Look at the labels on your clothes. These will show the proportions of different yarns in the fabric.

| 55% RAMIE |
| 45% COTTON |

$55\% = \frac{55}{100} = 0.55$
$45\% = \frac{45}{100} = 0.45$
$100\% \quad\quad\quad 1.00$

Material: 60% polyester, 50% viscose.

Washable. Material: jacket 44% viscose, 38% polyester, 18% linen.

| 100% COTTON |

shows that the fabric is made entirely from cotton

Material: 50% polyester, 50% viscose.

> The percentages on the labels always add up to 100%. The 100% refers to the whole garment.

Let's look more closely at that strange percentage sign, %.

In a way, the percentage sign is just 100 rearranged.

More mathematical shorthand!

This should help you to remember that a percentage is a fraction with a denominator of 100.

Per cent means 'in each hundred'. (That's Latin!).

23% means $\frac{23}{100}$.

An interesting fact

A electric eel can give a shock which is 64% of 625 volts.

That means it's $\frac{64}{100}$ of 625 volts.

> Of tells us to multiply.

$625 \times \frac{64}{100}$ volts = 400 volts

Keep away from electric eels – 400 volts is more than the voltage in your home!

Examples involving percentages

92% of a fruit drink is water. If you have 7 litres of the drink, how much is water?

We need to find 92% ($\frac{92}{100}$) of 7 litres

$$7 \times \frac{92}{100} \text{ litres} = 6.44 \text{ litres}$$

A garage increases the price of its unleaded petrol by 3% per gallon. The price was £1.99 per gallon before the increase. What is the new price?

We need to find 3% of £1.99

$$£1.99 \times \frac{3}{100} = £0.0597 = £0.06 \text{ to 2 d.p.}$$

The new price is £1.99 + the increase
= £1.99 + £0.06
= £2.05

A building society offers 95% mortgages to first-time buyers. How much would they offer on a house costing £64 000?

We need to find 95% of £64 000

$$£64\,000 \times \frac{95}{100} = £60\,800$$

79

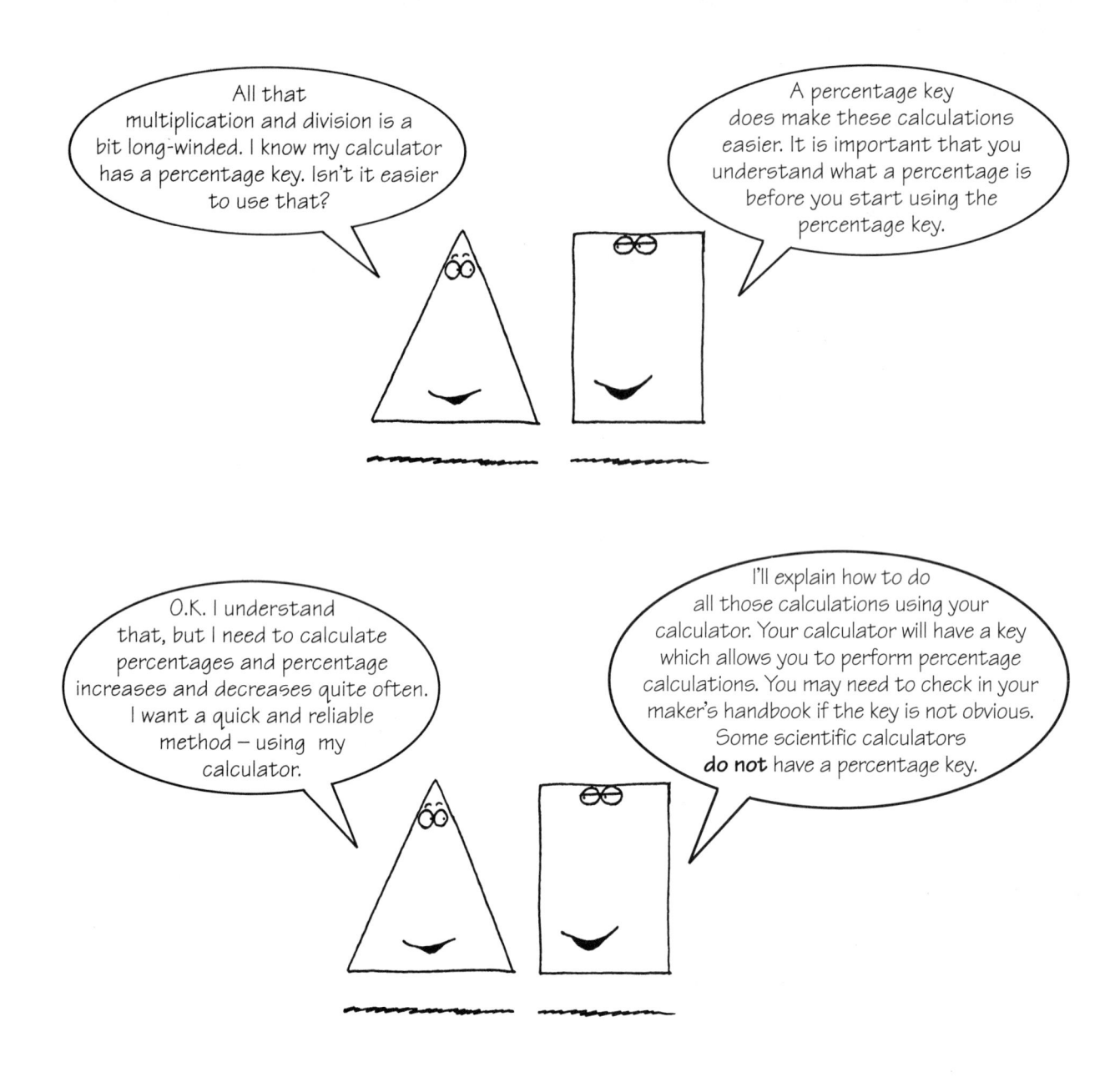

> All that multiplication and division is a bit long-winded. I know my calculator has a percentage key. Isn't it easier to use that?

> A percentage key does make these calculations easier. It is important that you understand what a percentage is before you start using the percentage key.

> O.K. I understand that, but I need to calculate percentages and percentage increases and decreases quite often. I want a quick and reliable method – using my calculator.

> I'll explain how to do all those calculations using your calculator. Your calculator will have a key which allows you to perform percentage calculations. You may need to check in your maker's handbook if the key is not obvious. Some scientific calculators **do not** have a percentage key.

Percentages on the calculator

If your calculator has a percentage key it will probably look like $\boxed{\%}$ or $\boxed{\Delta\%}$.

On most scientific calculators the percentage key is a shift or second function key. This means the symbol is not *on* a key but just *above* a key.

$$\text{Casio} \quad \overset{\%}{\boxed{=}} \qquad \text{Sharp} \quad \overset{\Delta\%}{\boxed{x^2}}$$

To use these shift keys you must press the Shift or Second function key first, after entering the figures.

Casio $\boxed{\text{SHIFT}}\ \boxed{\%}$ Sharp $\boxed{\text{2ndF}}\ \boxed{\Delta\%}$
(older models) $\boxed{\text{INV}}\ \boxed{\%}$

The percentage must come at the *end* of the calculation. You will not need to press the = key.

Find 8% of 50.

$$\boxed{5}\,\boxed{0}\,\boxed{\times}\,\boxed{8}\,\boxed{\text{SHIFT}}\,\boxed{\%} \quad 4$$

What is 250 as a percentage of 400?

$$\boxed{2}\,\boxed{5}\,\boxed{0}\,\boxed{\div}\,\boxed{4}\,\boxed{0}\,\boxed{0}\,\boxed{\text{SHIFT}}\,\boxed{\%} \quad 62.5$$

▽ Try some for yourself

8. Find the values of the following
   (a) 15% of £32.40
   (b) 10% of £159.65
   (c) 17.5% of £699.99
   (d) 26 as a percentage of 200
   (e) 33.3 as a percentage of 160

Percentage increase and decrease

All that is quite easy. I think the percentage key is a great help. Most of my tasks involve adding VAT or finding the cost of something excluding VAT. How do I deal with real situations like reductions in sales?

These are all situations involving percentage increase or decrease. They are the most frequent practical examples of percentages. We need to get to grips with them.

A leather briefcase costs £49.99 excluding VAT at 17.5%. What is the cost with the VAT added?

Total cost = £49.99 + (17.5% of £49.99) $£49.99 \times \dfrac{17.5}{100}$

$$= £8.748 \times 25$$

17.5% of £49.99 is   £8.748 25

= £8.75 to 2 d.p.

£49.99 to £8.75 = £58.74

The *actual* cost is £58.74.

Remember – most practical measures are rounded to 2 d.p. The extra decimal places have no meaning when dealing with money.

A leather jacket costs £129.99.

The price is reduced by 10% in the January Sales.

What is the sale price?

Sale price = £129.99 − 10% of £129.99
= £129.99 − £13.00
= £116.99

10% od £129.99

$£129.99 \times \dfrac{10}{100} = £12.999$

= £13.00 to 2 d.p.

The sale price is £116.99.

Some calculators will allow you to increase (or decrease) an amount by a given percentage. All you need to do is to press $\boxed{+}$ (or $\boxed{-}$) after the normal percentage calculations. *Not all calculators have this facility.* Check your handbook.

Increase 120 by 30%.

$\boxed{1}\boxed{2}\boxed{0}\boxed{\times}\boxed{3}\boxed{0}\boxed{\text{Shift}}\boxed{\%}\boxed{+}$ 156

Decrease 280 m by 18%.

$\boxed{2}\boxed{8}\boxed{0}\boxed{\times}\boxed{1}\boxed{8}\boxed{\text{Shift}}\boxed{\%}\boxed{-}$ 229.60 m

How much would you pay for a suit costing £110 if there is a discount of 15%?

$\boxed{1}\boxed{1}\boxed{0}\boxed{\times}\boxed{1}\boxed{5}\boxed{\text{Shift}}\boxed{\text{Shift}}\boxed{\%}\boxed{-}$ £93.50

▽ Try some for yourself

9. Find the price you pay if there is a discount of:

   (a) 5% on £80

   (b) 12.5% on £60

   (c) 25% on £134.60

   (d) 10% on £75

10. The Council Tax is to be increased by 7.2%. Find the increased rate when the amount paid in the current year is:

    (a) £368

    (b) £171

    (c) £257

    (d) £468.60

11. On a housing estate 20% of households have the Daily Planet delivered. 24% have the Daily News, 18% have The Ford Echo, and 12% have the Starkland Star. If the estate has 150 houses on it, how many households do not have a paper delivered?

12. (a) An electric oven costs £435.50 plus VAT at 17.5%. How much is the cost of the oven?

    (b) A cooker was bought for £435 and sold to make a profit of 7%. What was the selling price?

13. (a) A hospital has 240 beds, of which 210 are occupied. What percentage of the beds are unoccupied?

    (b) At a major ambulance depot 22 out of 30 ambulances are serviceable and the rest are off the road for maintenance or repair.

    What percentage of the total number of ambulances are not available for duty?

14. A holiday package offered by a travel agency cost £276.50 per person last year. The price for this year is to be increased by 5%.

    (a) What is the new price?

    (b) What will the holiday cost with the 10% discount?

What have you learnt about ratios, scales and percentages?

I am no longer frightened when faced with ratios, scales and percentages. I feel quite confident when handling them.

- ✓ I can write ratios in their lowest form.
- ✓ I can divide amounts into given ratios.
- ✓ I can scale recipes.
- ✓ I can interpret scales on diagrams and maps.
- ✓ I can change fractions and decimals to percentages.
- ✓ I can use my calculator to work out percentages, including percentage increases and decreases.
- ✓ I can solve problems involving ratios, scales and percentages.

Further Questions

1. The table shows the prices for a 7-day holiday in Orlando, Florida. The prices are for adults. There is a child reduction of 60% on all prices. Calculate the child prices for each hotel.

Hotel	1 July–31 August
Days Inn	£598
Marriott	£676
Travelodge	£730

   Use this information to find the cost for a family of two adults and two children to stay at each hotel.

2. A computer magazine lists all prices exclusive of VAT. To find out the true cost of each computer VAT at 17.5% must be added to the advertised price.

   Calculate the actual price of each of these computers.

   (a) 386 SX33   £585.00

   (b) 386 DX40   £655.00

   (c) 486 SX25   £790.00

   (d) 486 DX33   £990.00

3. A department store offers a 10% reduction on purchases made by store-card customers. What will be the savings on purchases totalling £47.03?

4. The ingredients for a cake have a total mass of 3.8 kg. If the amount of mixed fruit is 17% and the amount of flour is 42%, what is the mass, in kilograms of the remaining ingredients?

5. The current flowing through a circuit is increased from 7 amperes to 7.68 amperes. What is the percentage increase in current?

6. The ordinary price of a certain clutch is £50. Due to an increase in cost of the material the price is increased by 5%.

   (a) Find the new charge.
   A customer is allowed a discount of 25%.

   (b) Find the actual amount he pays.

7. A variation in weight of 2% either side of the standard weight is allowed in electric conductors. The standard weight of 1000 m of 37/13 conductor is 1320 kg. Find the maximum and minimum weights allowable.

8. The Admiralty test load on a $\frac{1}{2}$" chain crane is 3 tonnes. At the Marwick Works the test load is 10% higher. What is the Marwick test load?

9. A salt solution is made by adding 65 grams of salt to 250 grams of water. Find the percentage of salt in the total solution.

10. A commercial traveller receives as commission $2\frac{1}{2}$% on all his business over £500. In a certain month he receives orders for £3050. What commission does he receive?

11. A piece of land measuring $\frac{1}{2}$ acre is to be converted into a garden. Three-fourths of the area will be garden and the rest paving.

    What will be the ratio of paving to garden?

12. Three cousins inherited £32 818 from an uncle. The terms of the will dictate that the money shall be divided in the ratio $2:3:4$. How much does each cousin receive?

13. Concrete is made up of the following ingredients: 1 part cement, 3 parts sand, 6 parts aggregate. What quantities will be needed to make up 75 kg of concrete?

14. Men and women guests at a retirement dinner are in the ratio $7:5$. If there are 60 guests at the dinner, determine the number of men and women.

15. The ratio of incoming calls to outgoing calls at a hospital in a 5-day period was $11:4$. If the number of outgoing calls was 128, how many incoming calls were made?

16. A local building society has 900 customers holding chequeing accounts and 150 customers holding deposit accounts. What is the ratio of current accounts to deposit accounts.

17. A manufacturing company spends £300 000 on building alterations, £150 000 on new machinery and £50 000 on research. What is the ratio of these amounts?

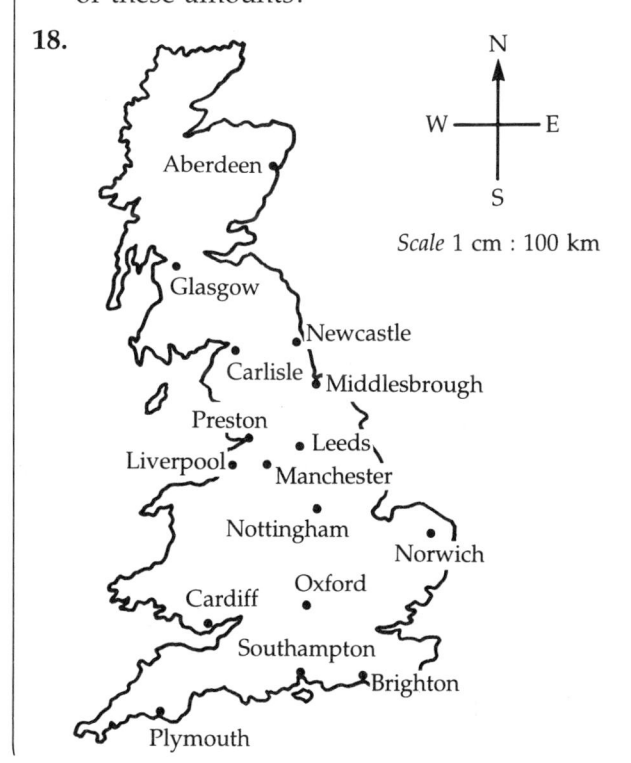

18.

Scale 1 cm : 100 km

85

In the following football matches, find out how far the away team has had to travel to the nearest 10 km.

(a) Norwich v Manchester City

(b) Newcastle v Liverpool

(c) Manchester United v Southampton

(d) Leeds v Norwich City

(e) Oxford United v Nottingham County

(f) Plymouth v Brighton

(g) Carlisle v Preston

(h) Glasgow v Aberdeen

(i) Middlesbrough v Nottingham Forest

(a) Manchester City travel to Norwich
Distance on the map = 2.5 cm
Actual distance = 2.5 × 100 km
= 250 km

Try the rest for yourself.

19. This scale drawing is of part of the foundations of a house.

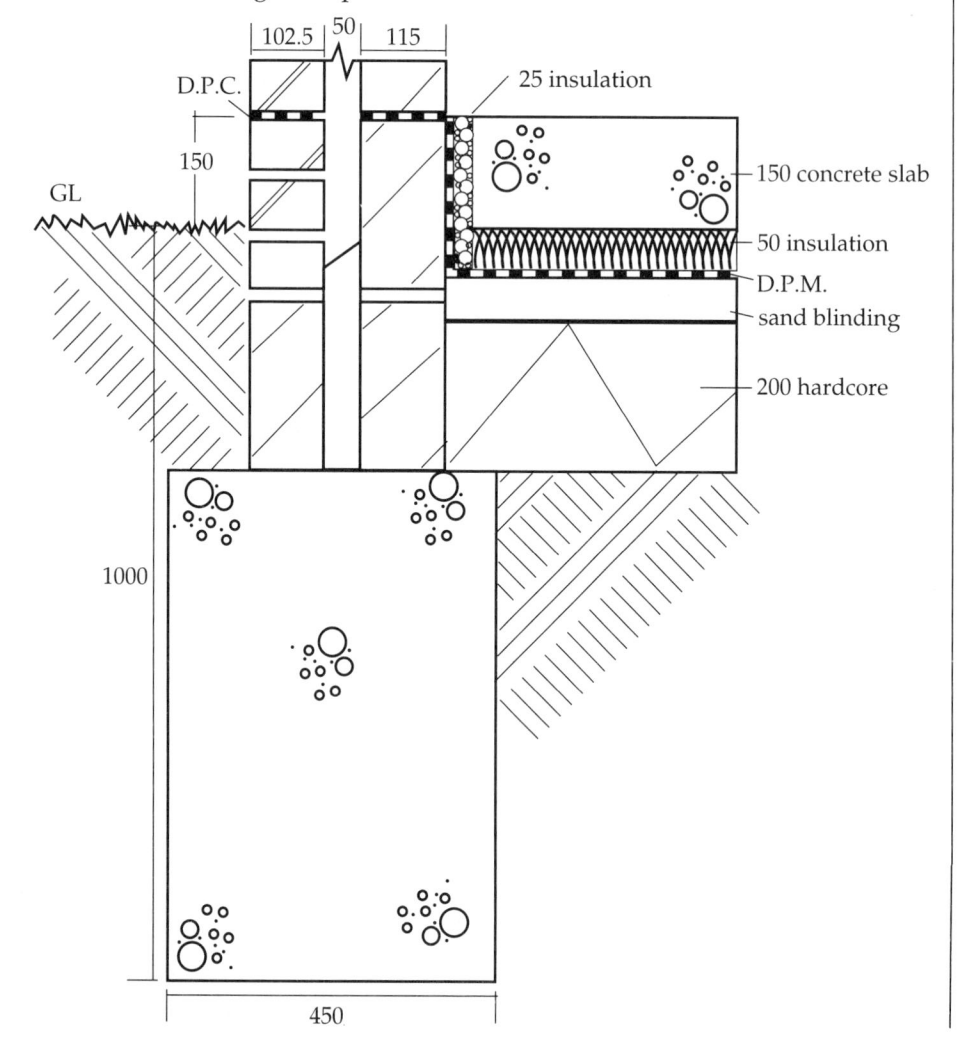

The actual measurements, in millimetres, are indicated on the plan.

   (a) What is the scale of the plan?

   (b) The actual depth is 1000 mm. What is the depth on the plan?

   (c) The width of the foundation is 450 mm. What is the width on the plan?

20. The main ingredients for French onion soup are

   500 g onions
   1.2 litres stock
   This makes 4 servings.

   What will be the quantities required for

   (a) 6 servings

   (b) 10 servings

   (c) 20 servings?

Solutions

1. (a) $5 : 3 : 2 = 10$ parts

$\frac{5}{10}$ of $1500 = 1500 \times \frac{5}{10} = 750$

$\frac{3}{10}$ of $1500 = 1500 \times \frac{3}{10} = 450$

$\frac{2}{10}$ of $1500 = 1500 \times \frac{2}{10} = 300$

(b) $8 : 2 = 10$ parts

male: $150 \times \frac{8}{10} = 120$

female: $150 \times \frac{2}{10} = 30$

2. (a) 13 cm   (b) 13×100 cm $= 13$ m

3. (a) 8 cm   (b) 8×100 cm $= 8$ m

4. (a) 5 cm $\times$ 4 cm represents 5 m $\times$ 4 m

(b) 4.7 cm $\times$ 4 cm represents 4.7 m $\times$ 4 m

(c) Bedroom 1 is larger.

5. $8\frac{1}{2}$ cm represents 5666.67 m
$= 5.67$ km

6. $1080 = 720 \times 1.5$
yeast: $20 \times 1.5 = 30$ g
flour: $700 \times 1.5 = 1050$ g

7. $6 : 4 = 10$ parts
lead: $2.5 \times \frac{6}{10} = 15$ g
tin: $25 \times \frac{4}{10} = 10$ g

8. (a) £4.86   (d) 13%

(b) £15.97   (e) 20.8%

(c) £122.50

9. (a) £76   (c) £100.95

(b) £52.50   (d) £67.50

10. (a) £394.50   (c) £275.40

(b) £184.38   (d) £502.34

11. $150 \times \frac{20}{100} = 30$

$150 \times 24\% = 36$
$150 \times 18\% = 27$
$150 \times 12\% = \underline{18}$

Total $= 111$

$150 - 111 = 39$ houses have no newspaper delivered

12. (a) £435.50 + VAT at 17.5%
VAT $=$ £76.21
Total cost $=$ £511.71

(b) 7% of £435 $=$ £30.45
Selling Price $=$ £465.45

13. (a) $240 - 210 = 30$ $\frac{30}{240} \times 100 = 12.5\%$

(b) $30 - 22 = 8$ $\frac{8}{30} \times 100 = 26.67\%$

14. (a) £276.50 + 5% $=$ £276.50 + £13.83
$=$ £290.33

(b) £290.33 − 10% discount $=$ £290.33 − £29.03
$=$ £261.30

87

6 Conversions

I have been asked to convert from metric to imperial measures, or the other way round, several times. I am not really sure how I am expected to do these conversions. Is there a correct way to do it?

Conversion, particularly between metric and imperial measures is quite a common requirement. There are three methods you should know about. You can use conversion tables, parallel scales or conversion graphs.

I might have known that the answer would not be simple? Why do I need to know three methods? I suppose they each have a particular use so I need to know how to use them all.

You are beginning to understand how application of number works. If you need to convert currency, temperatures or measurements you need to know the best method to use in each case.

Conversion tables

Tell me what you mean when you say that I can use a conversion table. What is a conversion table? Do I find it in a book or do I have to work it out for myself?

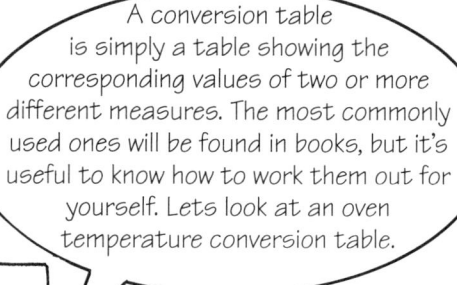

A conversion table is simply a table showing the corresponding values of two or more different measures. The most commonly used ones will be found in books, but it's useful to know how to work them out for yourself. Lets look at an oven temperature conversion table.

Oven temperatures

°C	°F	Gas Mark
110	225	$\frac{1}{4}$
130	250	$\frac{1}{2}$
140	275	1
150	300	2
170	325	3
180	350	4
190	375	5
200	400	6
220	425	7
230	450	8
240	475	9

A recipe for roast lamb gives an oven setting of Gas Mark 6. An electric oven will have the temperature given in °F or °C. Convert the setting to suit an electric oven.

$$\text{Gas Mark 6} = 200°C \text{ or } 400°F$$

▽ Try some for yourself

1. In each question give the equivalent oven settings.

   (a) A recipe for a rich chocolate cake gives an oven temperature of 180°C.

   $$180°C = \quad °F = \text{Gas Mark}$$

   (b) Lasagne is to be cooked at Gas Mark 5.

   $$\text{Gas Mark 5} = \quad °F = \quad °C$$

   (c) Pavlova needs a cool oven. The setting is given as 275°F.

   $$275°F = \quad °C = \text{Gas Mark}$$

That is such an obvious conversion table! I had never thought of it like that when I have been using recipes. That one is very simple, just three columns of corresponding oven settings. I've seen conversion charts for litres to gallons at the petrol station. Have you got one of those?

Here is a conversion table for gallons to litres and also one for miles to kilometres. The two go together as we often need to convert distance and petrol consumption. The tables are used by looking down the centre column and then reading across to the measure you want.

Volume

gal		litres
0.22	1	4.55
0.44	2	9.09
0.66	3	13.64
0.88	4	18.18
1.10	5	22.73
1.32	6	27.28
1.54	7	31.82
1.76	8	36.37
1.98	9	40.91

1 litre = 0.22 gallon
1 gallon = 4.55 litres

Distance

miles		km
0.62	1	1.61
1.24	2	3.22
1.86	3	4.83
2.49	4	6.44
3.11	5	8.05
3.73	6	9.66
4.35	7	11.27
4.97	8	12.87
5.59	9	14.48

1 mile = 1.61 km
1 km = 0.62 miles

▽ Try some for yourself

2. Convert from litres to gallons.

(a) 4 litres   (b) 6 litres   (c) 9 litres

3. Convert from kilometres to miles

(a) 4 km   (b) 6 km   (e) 9 km

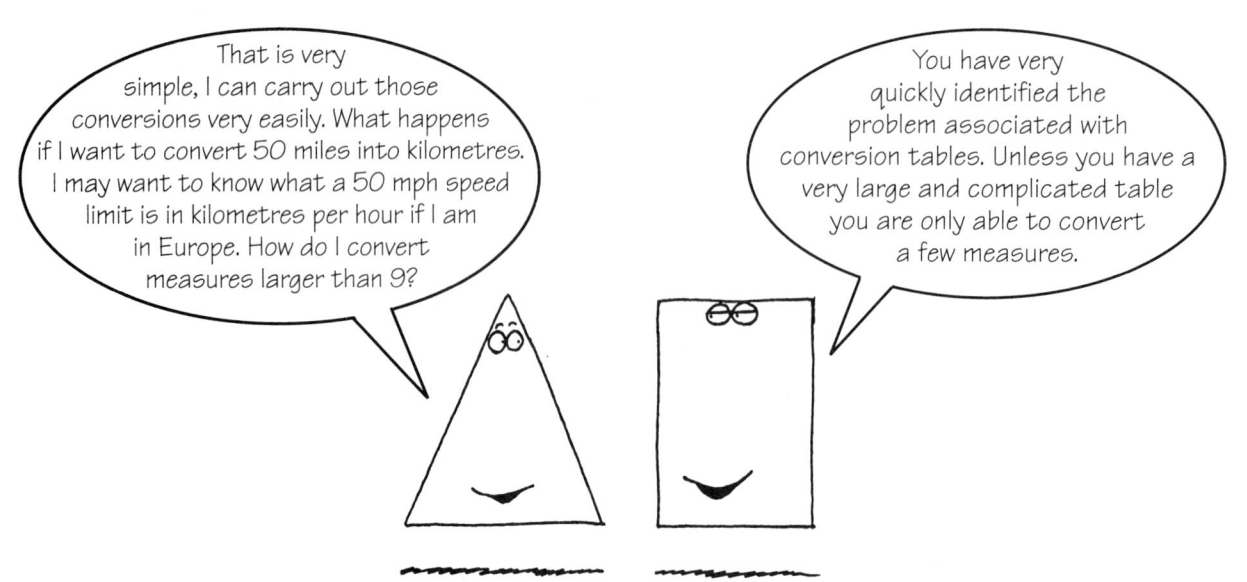

That is very simple, I can carry out those conversions very easily. What happens if I want to convert 50 miles into kilometres. I may want to know what a 50 mph speed limit is in kilometres per hour if I am in Europe. How do I convert measures larger than 9?

You have very quickly identified the problem associated with conversion tables. Unless you have a very large and complicated table you are only able to convert a few measures.

Conversion table: miles to kilometres

	0	1	2	3	4	5	6	7	8	9
0		1.61	3.22	4.83	6.44	8.05	9.66	11.27	12.87	14.48
10	16.1	17.71	19.32	20.93	22.54	24.15	25.76	27.37	28.98	30.59
20	32.2	33.81	35.42	37.03	38.64	40.25	41.86	43.47	45.08	46.69
30	48.3	49.91	51.52	53.13	54.75	56.35	57.96	59.57	61.18	62.79
40	64.4	66.01	67.62	69.23	70.84	72.45	74.06	75.67	77.28	78.89
50	80.5	82.11	83.72	85.33	86.94	88.55	90.16	91.77	93.38	94.99

Using the table,

50 miles = 80.5 km   25 miles = 40.25 km

So 50 mph = 80.5 kph   25 mph = 40.25 kph

▽ Try some for yourself

4. Convert to kph.

(a) 30 mph   (b) 40 mph   (c) 55 mph

5. Convert to mph.

(a) 45.08 kph   (b) 72.45 kph   (c) 94.99 kph

The conversion from mph to kph was easy. I found that converting from kph to mph was quite difficult. I had to find the numbers in the table and that was rather confusing. I really needed another table to convert from kilometres to miles. How could I construct a table like that?

All you need to construct your own conversion table is the basic conversion. One kilometre, = 0.62 miles. All you have to do then is multiply 0.62 by all the number of kilometres you want in your table. I'll start you off and you can finish the table.

Construct a table to convert kilometres to miles. The table must contain values from 0 to 30 kilometres.

Kilometres to miles

	0	1	2	3	4	5	6	7	8	9
0		0.62	1.24	1.86	2.49	3.11	3.73	4.35	4.97	5.59
10	6.2	6.82	7.44	8.06						
20										
30										

The first row of figures is taken from our original conversion table. The remaining conversions must be carried out in this way:

10 kilometres = 0.62 × 10 miles = 6.2 miles
11 kilometres = 0.62 × 11 miles = 6.82 miles
12 kilometres = 0.62 × 12 miles = 7.44 miles

Use your calculator's memory to make the calculations easier.

Store 0.62 in the memory.

$$\boxed{0}\ \boxed{.}\ \boxed{6}\ \boxed{2}\ \boxed{Min}$$

Multiply each number by the value in the memory.

$$\boxed{1}\ \boxed{3}\ \boxed{\times}\ \boxed{RM}\ \boxed{=}\quad 7.93$$
$$\boxed{MR}$$

▽ Try some for yourself

6. Complete the kilometres to miles conversion table.

Parallel scales

You said that I can use parallel scales for conversions. What are parallel scales? They sound like a cross between parallel lines and kitchen scales. This is maths so they must be complicated!

Actually, your suggestion is almost correct. A parallel scale has two parallel sets of measurements or scales. Parallel scales are another way of helping you to convert from one measurement to another.

What do parallel scales look like? I am sure I have never seen one. It must be something that is not used very often.

Some parallel scales are very commonly used. You just did not realise what they were. I shall show you two of the most commonly used parallel scales.

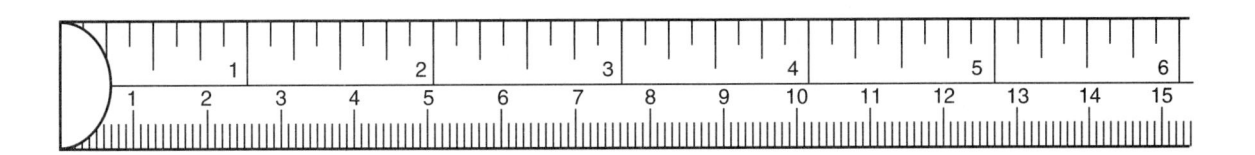

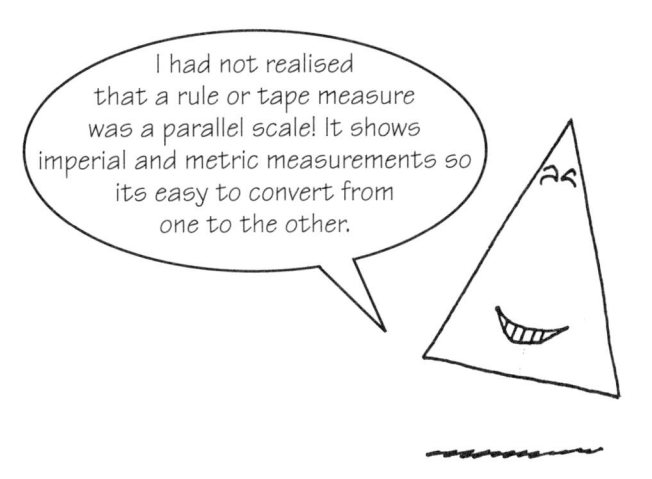

I had not realised that a rule or tape measure was a parallel scale! It shows imperial and metric measurements so its easy to convert from one to the other.

1 inch = 2.5 cm
5 cm = nearly 2 inches

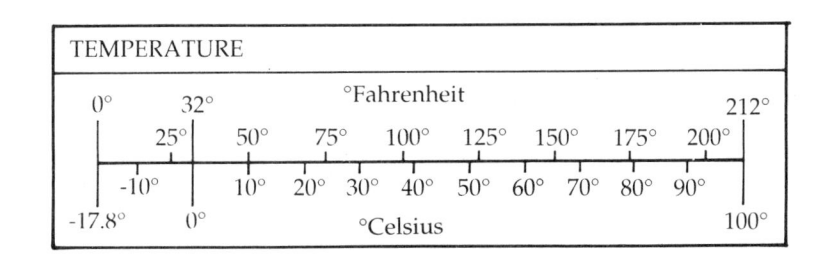

What is the other parallel scale you are going to show me? I expect I shall feel so stupid when I see it. A rule is so obvious. What is the other obvious scale?

This is the other commonly used parallel scale. The thermometer gives temperature in °C and °F. It's the parallel scale we use for converting temperature.

TEMPERATURE

°Fahrenheit: 0°, 32°, 25°, 50°, 75°, 100°, 125°, 150°, 175°, 200°, 212°

°Celsius: -17.8°, -10°, 0°, 10°, 20°, 30°, 40°, 50°, 60°, 70°, 80°, 90°, 100°

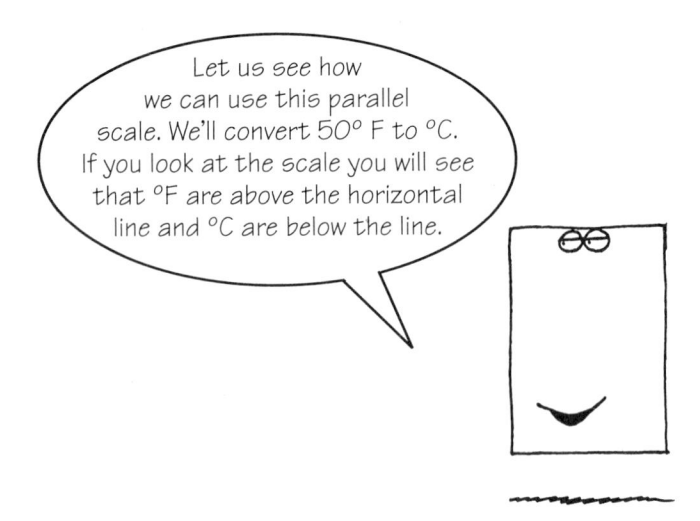

> Let us see how we can use this parallel scale. We'll convert 50° F to °C. If you look at the scale you will see that °F are above the horizontal line and °C are below the line.

If we find 50°F and then look *below* the line we shall find the corresponding temperature in °C. 50°F corresponds to 10°C.

> That seems quite easy because 50°F and 10°C are on the same vertical line. Not many of the numbers are like that. How would I convert 100°F to 0°C. There's nothing on the °C side of the line opposite to 100°F.

> I'm glad you asked that question. This is quite a problem with parallel scales. Let's see what we can do. 100°F lies between 30°C and 40°C. It's nearer to 40°C so we have to estimate its value – around 37°C or 38°C.

> But that is not a very accurate answer, is it? What would happen if I needed to be really accurate? I don't think these parallel scales will be good enough.

> You have identified the problem we have when using parallel scales. They are fine when we need only an approximate answer. If we need accuracy we must use another method.

▽ Try some for yourself

7. Use this scale to convert:

    (a) 75°F to °C

    (a) 50°C to °F

    (b) −10°C to °F

    (c) 176°F to °C

    (d) 0°C to °F

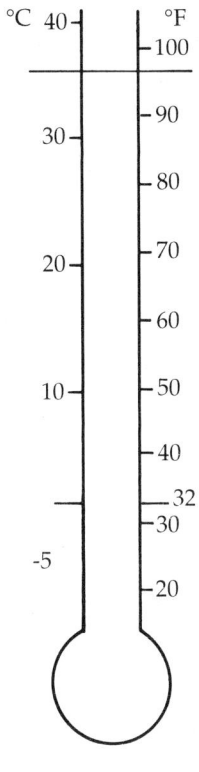

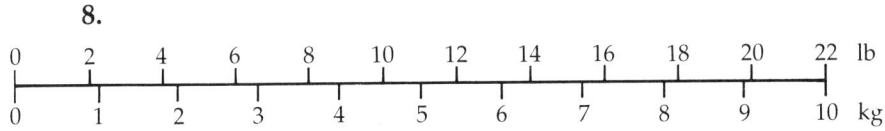

8.

Use this scale to convert:

    (a) 4 lb to kg   (d) 7 kg to lb

    (b) 10 kg to lb   (e) 1 kg to lb

    (c) 14 lb to kg   (f) 0 lb to kg

Conversion graphs

You have told me about conversion tables and parallel scales. We have found problems with both. You said that for accuracy we need another method of conversion. I suppose you mean we need to draw graphs?

Conversion graphs are a simple and relatively accurate method of conversion. They are easy to draw. They involve little calculation and they are easy to read.

95

> I think I remember drawing a graph to convert temperatures. It was quite easy to draw and I only needed a couple of values to draw the straight line.

> A conversion graph is easy to draw if you are given some values. I am afraid you need more than two pairs of values to make sure that your graph is correct. I always use three pairs of values. It's safer that way.

Plotting graphs

- Decide which value is to go on the horizontal axis and which on the vertical axis.
- Work out the scale you will need by looking at the largest value which has to be included. Try to choose a scale that will be easy to sub-divide.
- Draw the axes which must intersect at the origin for both sets of values. Origin = 0.
- Label the axes and divide according to the scale you have decided to use.
- Plot the points on the graph, marking each point with a small cross. To plot a point you find the horizontal value and then move vertically until you reach the vertical value.
- Join the points with a straight line.

Draw a conversion graph for °C to °F.

We need to plot three points to draw the graph.

°C	−18	0	10
°F	0	32	50

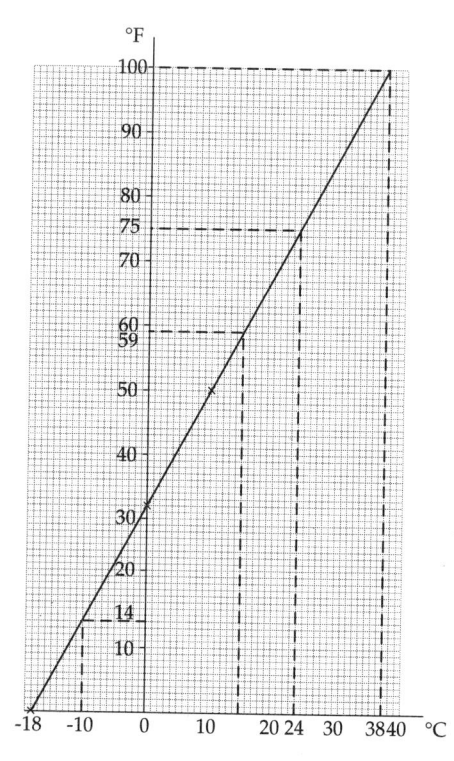

Convert 75 °F to °C.

Draw a horizontal line from 75 °F to the graph.

Draw a vertical line down to the °C axis.

Read off the value.

$$75°F = 24°C$$

Convert −10°C to °F.

Draw a vertical line from −10°C to the graph.

Draw a horizontal line to the °F axis.

Read off the value.
$$-10°C = 14°F$$

Convert 100°F to °C.
$$100°F = 38°C$$

Convert 15°C to °F
$$15°C = 59°F$$

> Converting temperatures by using that graph was really easy. It was much more complicated when we used the parallel scales. It is so easy to read values from a graph.

> It **is** easier to read values from a graph, it is also much more accurate. Some conversions cannot be displayed on a graph. The table of oven temperatures would not give a straight-line graph.

Currency conversion

Pounds to US dollars

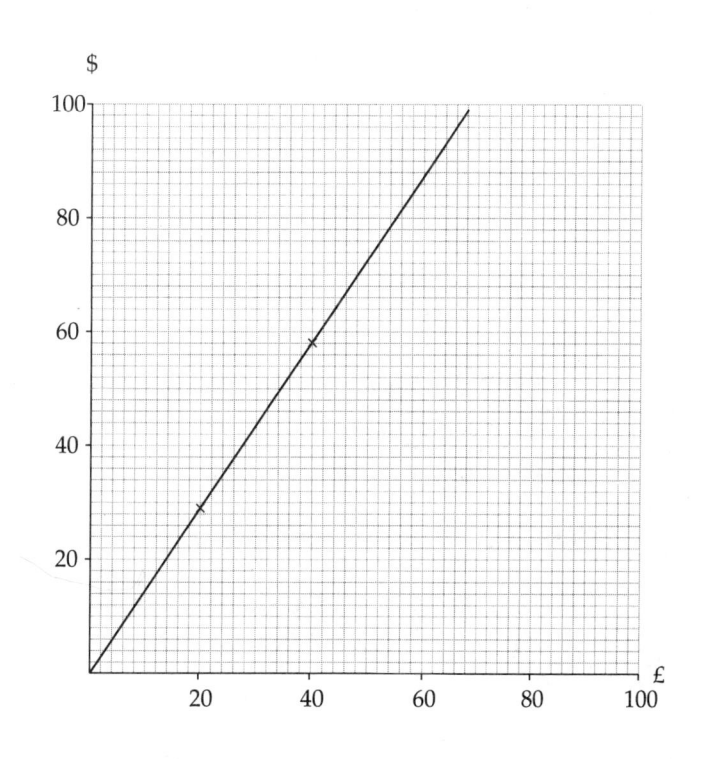

£1 = $1.45
£10 = $14.5
£20 = $29

▽ Try some for yourself

9. The exchange rate between the £ and the Australian dollar is £1 = $2. Construct a graph to show the value of the dollar in £s up to £100.

   From the graph find the value of:

   (a) £25 in Australian dollars

   (b) £75 in Australian dollars

   (c) $50 in £

   (d) $175 in £

10. The conversion from kilometres to miles is given as 1 km = 0.62 miles.

    Use this conversion to complete the table of values.

km	0	10	20
miles			

11. Construct a graph to allow you to convert from kilometres to miles up to 60 km.

    Find the metric equivalents of:

    (a) 25 miles

    (b) 30 miles

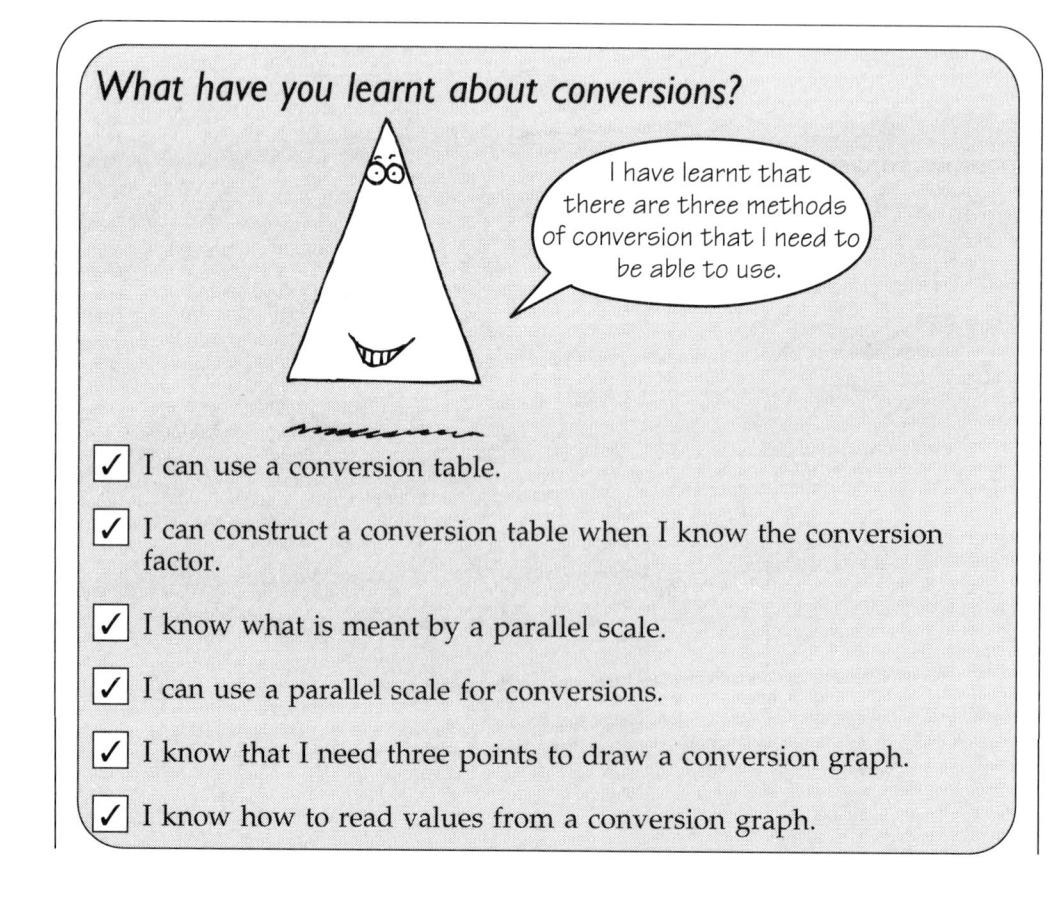

What have you learnt about conversions?

I have learnt that there are three methods of conversion that I need to be able to use.

- ✓ I can use a conversion table.
- ✓ I can construct a conversion table when I know the conversion factor.
- ✓ I know what is meant by a parallel scale.
- ✓ I can use a parallel scale for conversions.
- ✓ I know that I need three points to draw a conversion graph.
- ✓ I know how to read values from a conversion graph.

Further Questions

1. Use the parallel scale to complete the table.

   Average monthly temperatures

		Jan	Feb	Mar	Apr	May	June	July	Aug	Sept	Oct	Nov	Dec
Los Angeles	°F	65	66	65	67	69	72	75	77	77	74	70	66
	°C												
Orlando	°F												
	°C	22	23	25.5	28	31	33	33	33	32	28	25	22

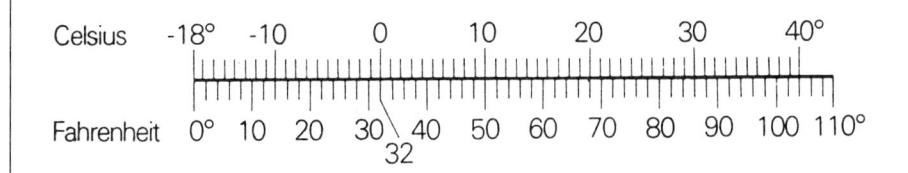

   Celsius -18° -10   0   10   20   30   40°

   Fahrenheit 0° 10 20 30 40 50 60 70 80 90 100 110°
                           32

2. Draw a graph using the table.

   Shoe sizes

British	6	7	8	9	10	11
Continental	$39\frac{1}{2}$	$40\frac{1}{2}$	$41\frac{1}{2}$	$42\frac{1}{2}$	$43\frac{1}{2}$	$44\frac{1}{2}$

   (a) Use the graph to convert the following British shoe sizes.

   (i) Size $6\frac{1}{2}$ (ii) Size 5   (iii) Size 3

   (b) Use the graph to convert the following Continental shoe sizes.

   (i) 37   (ii) $38\frac{1}{2}$ (iii) 39

3. Use the conversion 1 mile = 1.61 km to convert the figures for petrol consumption from miles per gallon to kilometres per gallon.

   $$10 \text{ mpg} = 10 \times 1.61 \text{ kpg} =$$

   (a) 15 mpg   (b) 20 mpg   (c) 30 mpg   (d) 35 mpg

4.   **Sizes of women's clothing**

British	10	12	14	16	18	20
Continental	40	42	44	46	48	50

   (a) Use the table to convert to continental sizes:

   (i) Dress: size 10   (ii) Blouse: size 14

   (b) Use the table to convert to British sizes:

   (i) Skirt: size 42   (ii) Coat: size 48

5. Use the table to convert the weights in the recipe to metric measure.

Weights

9 oz plain chocolate
6 oz butter
4 oz caster sugar
7 oz ground almonds

oz		g
0.04	1	28.35
0.07	2	56.70
0.11	3	85.05
0.14	4	113.40
0.18	5	141.75
0.21	6	170.10
0.25	7	198.45
0.28	8	226.80
0.32	9	255.15

6. Use the conversion 1 g = 0.04 oz to convert the metric measures given for the simnel cake.

175 g margarine
175 g sugar
225 g flour
75 g glacé cherries
50 g glacé pineapple

25 g glacé ginger
100 g currants
225 g sultanas
450 g marzipan

7.

Voltage (V)	0	1	2	3	4	5	6	7
Current (mA)	0	35	70	105	140	175	210	245

Draw a conversion graph.

Use your graph to find:

(a) The current for a voltage of 4.5 V.

(b) The voltage for a current of 200 mA.

8. (a) Use the table of values to draw a conversion graph.

Miles	0	1	5	10
Kilometres	0	1.61	8.05	16.1

(b) Use your graph to convert the following to kilometres.

(i) A running track ½ mile

(ii) A bus route 3.5 miles

(iii) Race track 1.6 miles

(iv) Sponsored walk 10 miles

(v) Newspaper round 2 miles

9. The exchange rate between the £ and German Mark is £1 = 2.5 Marks.

Draw up a table of values.

Draw a graph.

Use your graph to convert:

(a) £25 to Marks

(b) £100 to Marks

(c) 40 Marks to £

(d) 100 Marks to £

Solutions

1. (a) 180 °C = 350 °F = Gas Mark 4
   (b) Gas Mark 5 = 375 °F = 190 °C
   (c) 275 °F = 140 °C = Gas Mark 1

2. (a) 4 litres = 0.88 gall
   (b) 6 litres = 1.32 gall
   (b) 9 litres = 1.98 gall

3. (a) 4 km = 1.49 miles
   (b) 6 km = 3.73 miles
   (c) 9 km = 5.59 miles

4. (a) 30 miles = 48.3 km, 30 mph = 48.3 kph
   (b) 40 miles = 64.4 km, 40 mph = 64.4 kph
   (c) 55 miles = 88.55 km, 55 mph = 88.55 kph

5. (a) 45.08 km = 28 miles, 45.08 kph = 28 mph
   (b) 72.45 km = 45 miles, 72.45 kph = 45 mph
   (c) 94.99 km = 59 miles, 94.99 kph = 59 mph

6.

	0	1	2	3	4
0		0.62	1.24	1.86	2.49
10	6.2	6.82	7.44	8.06	8.68
20	12.4	13.02	13.64	14.26	14.88
30	18.6	19.22	19.84	20.46	21.08

	5	6	7	8	9
0	3.11	3.73	4.35	4.97	5.59
10	9.3	9.92	10.54	11.16	11.78
20	15.5	16.12	16.74	17.36	17.98
30	21.7	22.32	22.94	23.56	24.18

7. (a) 75 °F lies between 20 °C and 30 °C. It's nearer 20 °C so the answer will be between 23 °C and 25 °C.
   (b) 50 °C lies between 100 °F and 125 °F. It's very close to 125 °F so your answer should lie between 120 °F and 123 °F.
   (c) −10 °C lies between 0 °F and 15 °F. It's just over half-way, slightly nearer to 15 °F. Answer between 13 °F and 15 °F.
   (d) 176 °F lies between 80 °C and 90 °C but very close to 80 °C. 80 °C to 82 °C would give a reasonable answer.
   (e) 0 °C = 32 °F.

8. (a) 4 lb lies between 1 kg and 2 kg but much closer to 2 kg. Answer between 1.7 and 1.9 kg.
   (b) 10 kg lies just above 22 lb. Answer 22–22.2 lb.
   (c) 14 lb lies between 6 and 7 kg but closer to 6 kg. Answer between 6.1 and 6.4 kg.
   (d) 7 kg lies between 14 and 16 lb, but closer to 16 lb. Answer between 15.0 and 15.6 lb.
   (e) 1 kg lies between 2 and 4 lbs. It's very close to 2 lb. Most people know that 1 kg is equivalent to 2.2 lb.
   (f) 0 lb = 0 kg. Weightless is weightless in lb or kg.

9. (a) £25 = 50 dollars
   (b) £75 = 150 dollars
   (c) 50 dollars = £25
   (d) 175 dollars = £87.50

10.

km	0	10	20
miles	0	6.2	12.4

11. (a) 25 miles = 40 km
    (b) 30 miles = 48 km

101

7 Surveys and Questionnaires

Mori survey suggests

HOTEL OCCUPANCY SURVEY 1986

ON TRIAL JUICERS

A REVEALING NEW POLL

Surveys

The results of statistical surveys are widely available. The government is the largest producer of statistics in the country. The results of its surveys are available in the publications of Her Majesty's Stationery Office. Newspapers carry out surveys on a wide variety of topics and every week you can find the results in papers and magazines. We are surrounded by the results of statistical surveys many of which are of no great interest to the general population.

I am being asked to carry out surveys in different modules of my course. I did some at school but I think I need to know exactly what is involved.

You will be expected to carry out surveys as part of your GNVQ work. Sometimes you will collect the data by observation, sometimes by conducting interviews and sometimes by using questionnaires.

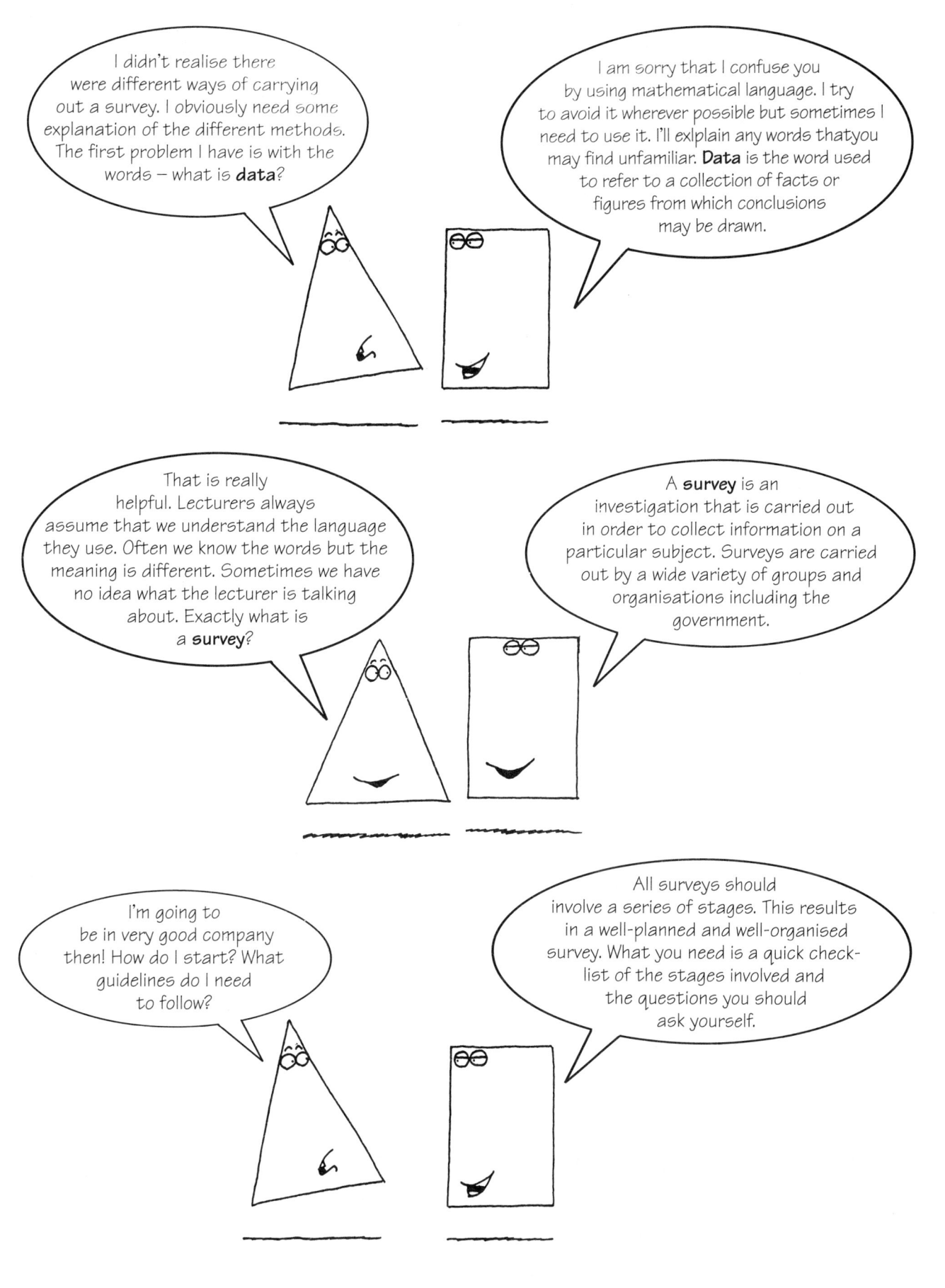

I didn't realise there were different ways of carrying out a survey. I obviously need some explanation of the different methods. The first problem I have is with the words – what is data?

I am sorry that I confuse you by using mathematical language. I try to avoid it wherever possible but sometimes I need to use it. I'll exlplain any words that you may find unfamiliar. Data is the word used to refer to a collection of facts or figures from which conclusions may be drawn.

That is really helpful. Lecturers always assume that we understand the language they use. Often we know the words but the meaning is different. Sometimes we have no idea what the lecturer is talking about. Exactly what is a survey?

A survey is an investigation that is carried out in order to collect information on a particular subject. Surveys are carried out by a wide variety of groups and organisations including the government.

I'm going to be in very good company then! How do I start? What guidelines do I need to follow?

All surveys should involve a series of stages. This results in a well-planned and well-organised survey. What you need is a quick check-list of the stages involved and the questions you should ask yourself.

That sounds very useful. Will this checklist apply to any type of survey?

Yes, and it can be used when you are constructing a questionnaire. I've broken it down into seven stages, beginning with the initial design and ending with the final product, the report.

Stages in carrying out a survey

Designing

How large shall I make it?
What type of sample shall I take?
How shall I collect the data?

Testing

How can I check the wording of my questions?
How can I check that the questions make sense?
Can I find two or three friends to try out the questions?

Collecting

How shall I collect the data?
Is it data that I can collect by observation?
Will an interview be the best method?
Should I design and use a questionnaire?

Organising

How shall I organise the questions?
How can I make sure the answers are easy to process?

Tabulating

How shall I group the data?
What classifications shall I use?
How can I arrange the data into tables?

Calculating

Do I need to work out percentages?
Do I need to find means, medians and modes?
Do I need to know what the ranges are?

Reporting

How shall I organise the data into a report?
What tables shall I include?
What graphs and diagrams will best illustrate my data?
What conclusions can I draw?
Can I explain reasons for my conclusions?

That seems quite straightforward. I shall need some help later when I have to tabulate, calculate and report. Now I need to know more about the three survey methods you mentioned.

I told you that you can collect data by three different methods. You can observe, you can interview or you can use a questionnaire. Let's look at the first method, collecting data by observation.

Surveys by observation

What do you mean when you say that data can be collected by observation? How can just looking provide me with information?

Think of the word observation. What does it make you think of? What do you associate with that word?

Observation? That makes me think of observatories and looking at the stars. Astronomers observe the heavens.

That's right. The observations made by astronomers result in the gathering of information about the stars and the planets. Astronomers conduct surveys by observation.

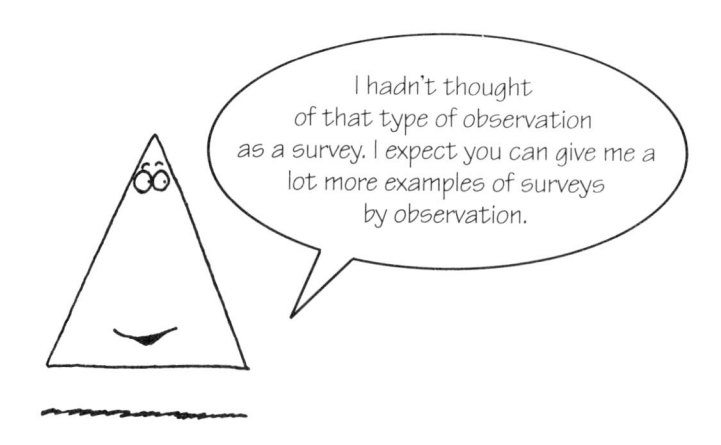

> I hadn't thought of that type of observation as a survey. I expect you can give me a lot more examples of surveys by observation.

Examples of surveys by observation

Sciences
Most scientific discoveries are the result of observation and experiment.

Social Sciences
Social scientists base their work on their observations of people and their behaviour.

Consumer surveys
Price comparison between stones.
Comparison of facilities e.g. for the disabled.
Testing manufactured products to find best buys, etc.
Tasting food and drink to find the best in a particular category.
Testing parts on a production line.

Surveys you could carry out

Observing the use of college car parks.
Observing the use of the library – peak times, etc.
Observing the use of college catering facilities.
Observing the use of college computer facilities.
Investigating facilities for the disabled.

Surveys by interview

Attitudes or opinions

The most common surveys of this type are opinion polls. These are commissioned by interested parties to find out what people think about certain questions. Political parties use opinion polls to measure their popularity. Newspapers use them to find out what their readers think about current news stories.

Reasons or motives

These surveys are used to find out why people do certain things. The Department of Transport carries out road surveys to find out where people are travelling and the reason for their journey. Data from such a survey can be used to plan new roads or simply the installation of traffic lights to control rush hour traffic.

Selection

Interviews are used to assess the suitability of a candidate for a place on a college course or for a job. In this case the interview is quite formal and has a definite purpose.

Reporting

Interviews are used by reporters in the news media seeking information leading to a news story.

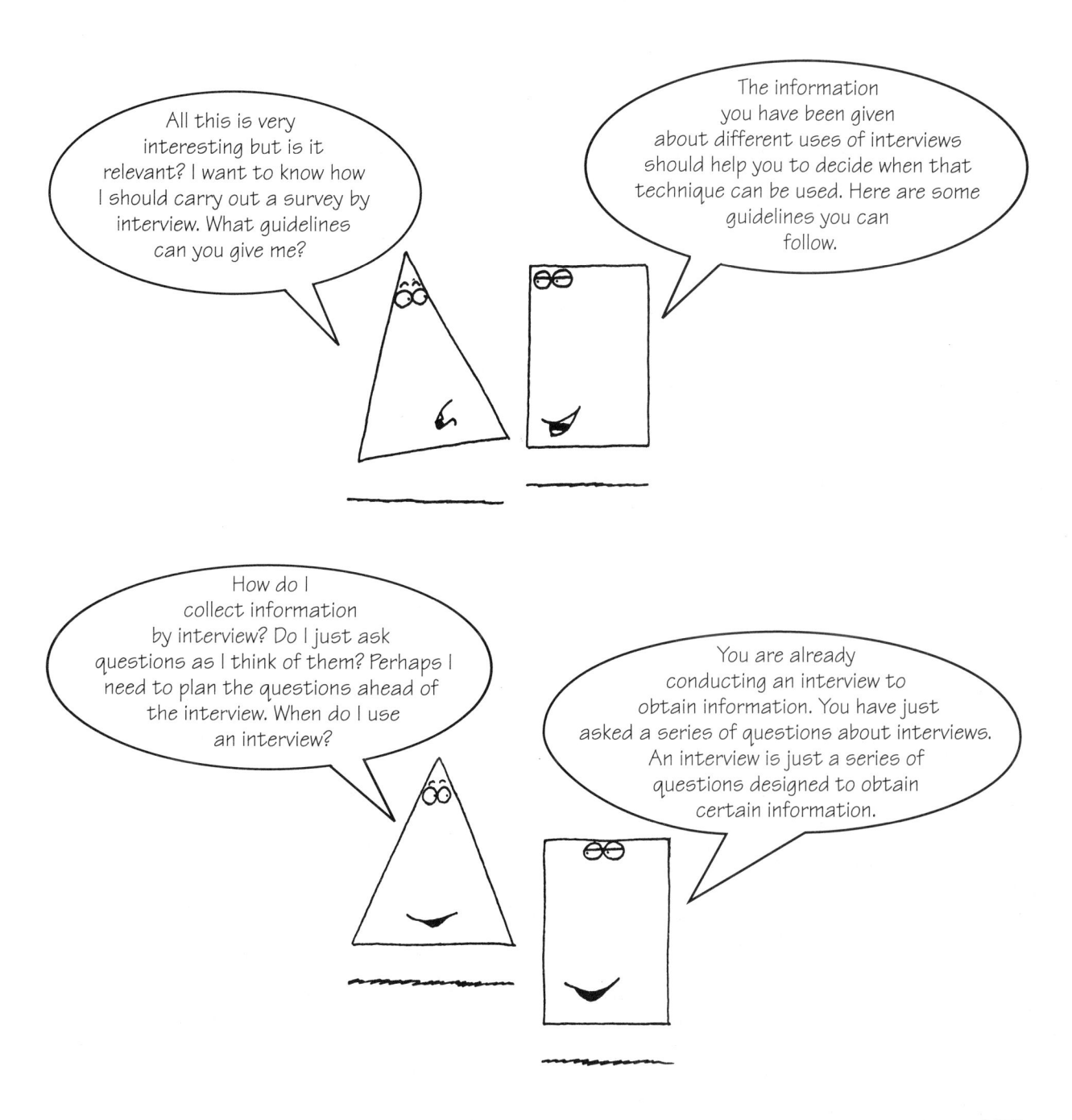

> All this is very interesting but is it relevant? I want to know how I should carry out a survey by interview. What guidelines can you give me?

> The information you have been given about different uses of interviews should help you to decide when that technique can be used. Here are some guidelines you can follow.

> How do I collect information by interview? Do I just ask questions as I think of them? Perhaps I need to plan the questions ahead of the interview. When do I use an interview?

> You are already conducting an interview to obtain information. You have just asked a series of questions about interviews. An interview is just a series of questions designed to obtain certain information.

*I suppose **I am** conducting an interview! I had not thought that that is what I do when I ask questions. I do it all the time! I'm an interviewer!*

When you conduct a survey by interview you need to plan your questions carefully. Interviews are used to collect information about attitudes and motives. I'll give you some examples.

Guidelines for interviews

- Decide what information you are seeking.
- Plan the questions to bring out that information.
- Try to arrange the questions in a logical order.
- Don't try to predict the answers by type-casting the person you are interviewing.
- Don't ask questions on a topic of which your respondent has no knowledge.
- Take care to record the answers accurately.
- Make sure that the results of the interview are reported accurately.
- Don't be tempted to make up answers yourself. Fictitious data serves no purpose.

Surveys by questionnaires

Interviews are quite frightening. I don't like approaching strangers and asking them questions. I get nervous and I'm sure they think I am a nuisance. I think I shall be happier with questionnaires.

Questionnaires are less daunting when you are not confident about approaching strangers. The problem with questionnaires is that not all are returned to you. The best way to be certain of getting them back is to deliver and collect them yourself — but that means you still have to speak to strangers.

> I think I could cope with that. My problem will be to produce a good questionnaire. I'm sure it's rather difficult to write a good questionnaire. Can you give me some guidelines?

> You are quite right, it is difficult to produce a good questionnaire. You can only do your best. I'll give you some guidelines to follow. They should help you to avoid the worst errors.

Guidelines for questionnaires

- Begin with the purpose of your survey. Explain what you are trying to do.

Example

> ### Attitudes to Learning Mathematics
> I am carrying out a survey into attitudes to learning mathematics among students on GNVQ courses. As part of the survey, please could you complete this questionnaire.

- Keep the questionnaire and questions short. Remember you are asking people to spend their time answering your questions.
- Make it clear where the answers are to be written and allow enough space for the answers.
- Set out the questions in a logical order so that the person answering can see how they follow on.
- Make sure that the questions are useful and relevant. They should be designed to produce the information you are looking for.
- Make sure that the wording of the questions is clear – avoid long, technical or unusual words.
- Avoid questions which may be too personal – some people are unwilling to give their names or age to a stranger.
- Try to avoid leading questions which suggest an answer. 'Don't you agree ...?' expects the answer 'yes'.
- Where possible use questions which require a 'yes' or 'no' answer or where a tick in a box indicates the answer.

(Please tick as appropriate ✓)

Do you have a part-time job ☐ Yes
☐ No

Male ☐
Female ☐

Are our staff helpful and friendly? ☐ Always
☐ Nearly Always
☐ Sometimes
☐ Hardly Ever
☐ Never

I was served in an acceptable
length of time ☐ Strongly Agree
☐ Tend to Agree
☐ Tend to Disagreee
☐ Strongly Disagree
☐ No Recent Experience

What age are you? ☐ 16–20
☐ 21–25
☐ 26–30
☐ 31–35
☐ 36–40
☐ 41–45
☐ 46 or over

What improvements would encourage you to use the canteen?
  (please state)

....

....

....

Thank you for completing this questionnaire.

What have you learnt about surveys and questionnaires?

I had no idea how to begin to carry out a survey. Now I know three ways of conducting one. I also know a little about writing questionnaires.

- ✓ I know the stages I should go through when I am asked to conduct a survey.
- ✓ I know that I can conduct a survey by observation.
- ✓ I know that I can conduct a survey by interview.
- ✓ I know the rules for writing questions for use in an interview.
- ✓ I know how to plan the questions for a questionnaire.

Further Questions

1. An FE college is located at a distance from the nearest bank. A group of GNVQ students have suggested that there is a need for a bank cashpoint to be installed on Campus. They decide to carry out a survey to see what the demand would be for such a facility. They need to know a number of facts:

   Which bank is used by the majority of students

   What facilities the students use: withdrawal,
                                      balance,
                                      statements, etc.

   How often students use their bank.

   What type of accounts students have: 14–18 yr,
                                          full accounts

   Write suitable questions to be included in the questionnaire.

2. The local lazer-game facility has noticed a fall in attendance. The manager wants to find out why young people are not using the facility.

   He asks this question: Why did you leave?

   (a) Explain why this is not a suitable question.

   (b) Suggest ways in which the question might be improved.

3. Comment on these following questions, explaining what problems will be experienced by people who try to answer them.

   (a) When did you start to arrive at lectures on time?

   (b) How many people are there in your family?

   (c) How many rooms are there in your house?

   (d) List three things you consider are bad for your health.

8 Tabulating your Information

I've carried out my survey – what do I do now? The list of stages puts tabulating after the survey. I'm not too sure what that means. Is it something to do with tables?

Tabulation means putting data into tables. After you have carried out your survey you have a lot of information but it's not very easy to extract, is it?

No. I have sheets of paper with answers to questions but I can't see the wood for the trees! What I need to do is to collect together, and count, all the answers to each question.

This is why you need to construct tables. Tables allow you to show your results in a clear, orderly way. They also help you to summarise the data and make it easier for you to see the pattern of your results.

How do I construct these tables? I expect there are certain rules that I should follow. Can you give me some advice? Perhaps you have another useful set of guidelines.

Tables

Guidelines for constructing tables

- All your tables must have a title.
- You must include the source of your data (usually below the table).
- Make sure your columns and rows have clear headings.
- Show clearly the units of measurement you are using.
- It's usually better to produce several simple tables than one very large and complicated one.
- If you include calculated figures, such as percentages, make sure that they are next to the figures from which they were derived.

Most of that is common sense and I'll try to stick to the guidelines. I would not have thought of including the source of my data. I suppose that tells the reader where to go to check on my figures.

That is one reason. It is also polite to acknowledge the source of your information. Of course, if it is the result of your own survey, then you indicate that at the bottom of the table.

We are surrounded by information presented in tables.

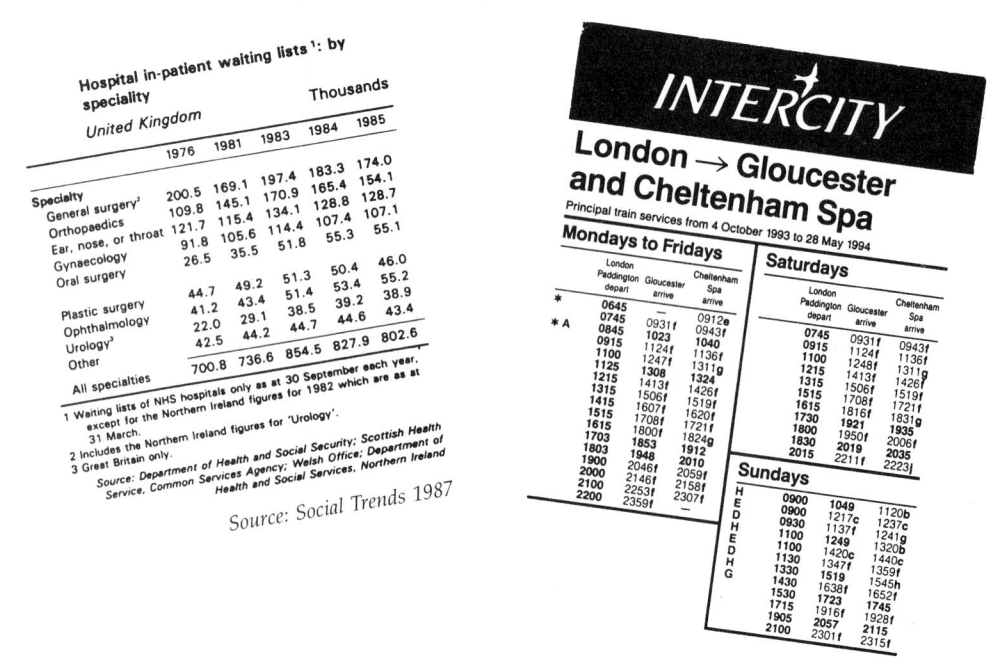

FLIGHT SAVERS

FROM	TO	DATE	PRICES	DAYS
LGW	MALAGA	DEC 12	£49	7
MAN	ARRECIFE	DEC 9	£49	7
LGW	MALTA	DEC 7,14	£49	7
LGW/MAN	FARO	DEC 11,12	£49	7
LGW/BHX	TENERIFE	DEC 7,10	£49	7
MAN	LAS PALMAS	DEC 11,13	£49	7
LGW/BHX	ARRECIFE	DEC 9	£49	7
LGW	MALAGA	DEC 11	£59	15
LGW	TENERIFE	DEC 14	£59	7
LGW	MALAGA	DEC 12	£59	14
LGW/MAN	TENERIFE	DEC 7	£69	14
MAN	FARO	DEC 12	£69	7
MAN	TENERIFE	DEC 14	£69	7
LGW	FUETEVENTURA	DEC 8	£69	7
LGW/MAN	MALAGA	DEC 19	£79	7
LGW/MAN	ALICANTE	DEC 21	£119	7
LGW/MAN	FARO	DEC 18,19	£119	7
LGW/BHX	LAS PALMAS	DEC 20	£139	7
LGW	ORLANDO	DEC 11	£219	7

ALL FLIGHTS SUBJECT TO AVAILABILITY.

EXAMPLES OF PERSONAL LOAN REPAYMENT TERMS
Secured Loans £3,000 - £15,000
Monthly Interest Rate 1.095%. Administration Fee £50

Amount of Loan £	Typical APR %	Repayment Term Months	With Payment Protection Plan — Monthly Payment £	With Payment Protection Plan — Total Payable over the term of the loan £	Without Payment Protection Plan — Monthly Payment £	Without Payment Protection Plan — Total Payable over the term of the loan £
3,000	14.8	60	75.31	4,568.60	68.47	4,158.20
	14.4	120	49.54	5,994.80	45.04	5,454.80
	14.2	240	38.99	9,407.60	35.45	8,558.00
5,000	14.4	60	125.52	7,581.20	114.12	6,897.20
	14.2	120	82.57	9,958.40	75.07	9,058.40
	14.1	240	64.98	15,645.20	59.08	14,229.20
7,500	14.2	60	188.29	11,347.40	171.19	10,321.40
	14.1	120	123.85	14,912.00	112.60	13,562.00
	14.0	240	97.47	23,442.80	88.62	21,318.80
10,000	14.2	60	251.05	15,113.00	228.25	13,745.00
	14.1	120	165.14	19,866.80	150.14	18,066.80
	14.0	240	129.96	31,240.40	118.16	28,408.40
12,500	14.1	60	313.81	18,878.60	285.31	17,168.60
	14.0	120	206.42	24,820.40	187.67	22,570.40
	14.0	240	162.46	39,040.40	147.70	35,498.00
15,000	14.1	60	376.57	22,644.20	342.37	20,592.20
	14.0	120	247.71	29,775.20	225.21	27,075.20
	14.0	240	194.94	46,835.60	177.23	42,585.20

Source: Halifax Building Society

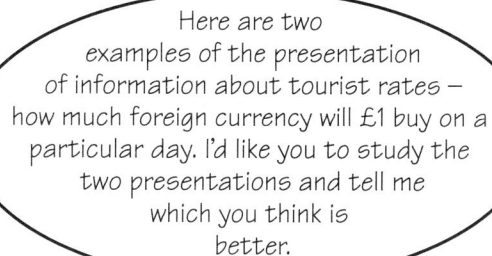

> Here are two examples of the presentation of information about tourist rates — how much foreign currency will £1 buy on a particular day. I'd like you to study the two presentations and tell me which you think is better.

DIVISION THREE

		Home						Away					
	P	W	D	L	F	A	W	D	L	F	A	Pts
Preston	17	7	0	2	24	12	4	2	2	17	15	35
Crewe	17	8	1	0	22	8	3	3	2	11	8	34
Walsall	17	4	3	2	16	11	5	2	1	8	4	32
Wycombe	17	4	2	2	15	11	4	5	0	16	12	31
Chester	17	4	4	1	12	9	3	3	3	8	7	27
Scunthorpe	17	4	3	1	17	6	3	3	3	18	15	26
Shrewsbury	17	3	3	2	10	9	4	2	4	15	7	11	26
Doncaster	17	5	1	2	13	9	3	4	2	12	11	26
Rochdale	17	2	2	16	6	2	2	4	12	12	25
Lincoln	17	4	2	3	14	15	3	2	3	13	15	25
Torquay	17	2	7	0	17	14	3	3	2	12	13	24
Colchester	17	6	1	2	20	15	1	1	6	13	25	23
Chesterfield	17	5	1	3	13	9	2	1	5	10	20	23
Bury	17	5	2	1	19	9	1	2	6	9	18	22
Mansfield	17	4	3	1	11	12	2	3	4	10	17	22
Gillingham	17	4	1	4	14	9	1	2	5	7	14	21
Carlisle	17	4	4	1	11	6	1	1	5	9	17	19
Scarborough	17	2	3	2	12	11	2	4	3	9	17	16
Wigan	17	2	4	16	18	2	2	5	10	23	15
Hereford	17	3	1	5	17	17	1	2	5	9	17	16
Darlington	17	2	4	2	8	12	1	0	2	6	7	18	11
Northampton	17	2	3	3	10	11	0	2	7	8	23	11

TOURIST RATES

TODAY'S tourist rates for sterling. £1 buys: Australia 2.11 dollars; Austria 17.80 schillings; Belgium 53.0 francs; Canada 1.92 dollars; Cyprus 0.76 pounds; Denmark 9.90 kroner; Finland 8.41 marks; France 8.62 francs; Germany 2.56 marks; Greece 373 drachmae; Holland 2.86 guilders; Hong Kong 11.25 dollars; Ireland 1.02 punts; Italy 2500 lira; Japan 164 yen; Malta 0.58 liri; New Zealand 2.60 dollars; Norway 11.0 kroner; Portugal 257 escudos; Spain 208 pesetas; Sweden 12.0 kronor; Switzerland 2.14 francs; Turkey 22075 lira; US 1.47 dollars.

Source: Gloucestershire Echo

TOURIST RATES

£1 Buys			
Australia(Dollars)	2.1100	Ireland(Punts)	1.0300
Austria(Schillings)	17.7500	Italy(Lira)	2500.0000
Belgium(Francs)	53.0000	Japan(Yen)	165.0000
Canada(Dollars)	1.9200	Malta(Liri)	0.5750
Cyprus(Pounds)	0.7650	New Zealand(Dollars)	2.6200
Denmark(Kroner)	9.9000	Norway(Kroner)	11.0400
Finland(Marks)	8.4700	Portugal(Escudos)	257.0000
France(Francs)	8.6000	Spain(Pesetas)	210.0000
Germany(Marks)	2.5600	Sweden(Kronor)	12.1000
Greece(Drachmei)	371.0000	Switzerland(Francs)	2.1500
Holland(Guilders)	2.8700	Turkey(Lira)	21773.0000
Hong Kong(Dollars)	11.3400	United States(Dollars)	1.4600

Source: The Independent

> There is really no comparison. The table of rates is clear and so easy to read. The other one is just a string of countries and strange words. It's very difficult to read. I suppose the table takes up more space but it has to be worth using the space.

> I just wanted you to see how much clearer information can become when it is put into a table. Now we shall find out how to tabulate our information.

Tally charts and frequency tables

Hotels in the West Midlands

Category	Rooms	Suites	4-Poster	Swimming	Price per Person
C	252	1		✓	£49
P	192			✓	£49
H	42	2			£49
P	136				£45
H	57	3			£49
H	118	4			£54
G	74		2		£79
H	52		5	✓	£54
H	67		1		£43
P	184				£45
C	147			✓	£49
C	122	1		✓	£54
H	87	3			£58
H	53				£49
H	34	16			£64
H	55			✓	£64
P	122			✓	£45
P	88			✓	£45
C	130				£43
H	40		1		£58
H	58	1			£54
H	20		2		£58
G	104	4	2		£74
P	60				£45
H	63	5	3		£74
H	37		2		£59
H	47		1		£54
H	29		2		£54

Source: Forte Leisure Breaks

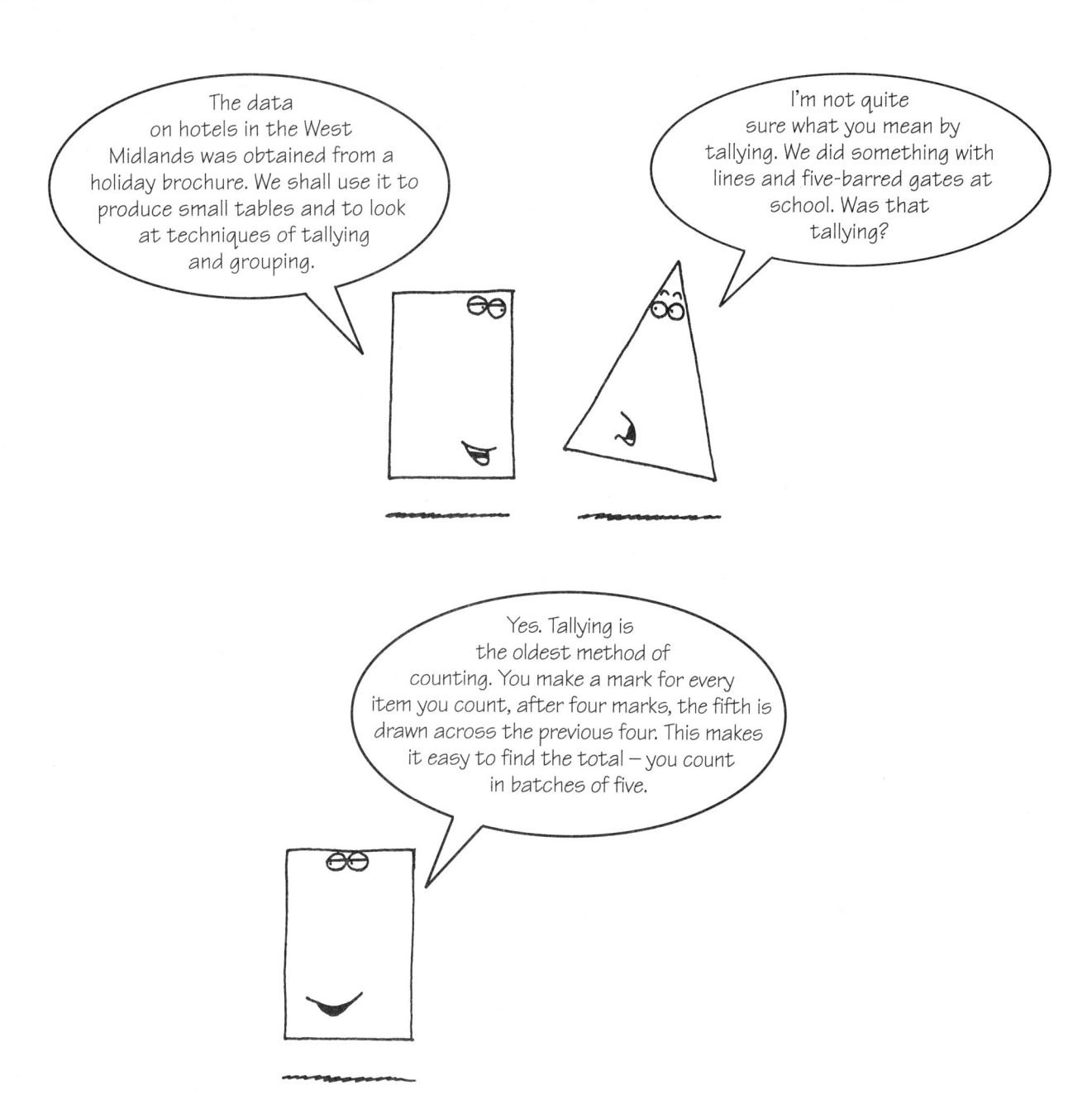

> The data on hotels in the West Midlands was obtained from a holiday brochure. We shall use it to produce small tables and to look at techniques of tallying and grouping.

> I'm not quite sure what you mean by tallying. We did something with lines and five-barred gates at school. Was that tallying?

> Yes. Tallying is the oldest method of counting. You make a mark for every item you count, after four marks, the fifth is drawn across the previous four. This makes it easy to find the total – you count in batches of five.

Hotels in the West Midlands by category

Category	Tally	Total
C	1111	4
G	11	2
H	1111 1111 1111 1	16
P	1111 1	6

The totals are called the frequencies. That tells us how often (or frequently) a category appears.

Looking at the table shows us that the most frequent category of hotel is category H. There are far more hotels in this category than in any other.

117

Try some for yourself

1. Use the data from the table about hotels in the West Midlands to complete this tally chart and frequency table.

Number of four-poster beds	Tally	Frequency
1		
2		
3		
4		
5		

2. (a) How many hotels are included in the survey?

(b) What percentage of these hotels offer four-poster beds?

(c) What percentage do not offer four-poster beds?

Grouped frequency tables

If we look at the data on the number of rooms we can see that the number varies from 29 to 252. This range is too large to list all the individual numbers. We need to group numbers together to compile a grouped frequency table.

We want to have between 5 and 10 groups, more than 10 becomes very messy.

We could have group of 50:

0–49, 50–99, 100–149, . . . , 250–299.

The grouped frequency table will look like this:

Number of rooms	Tally	Frequency
0–49	ⅢⅠ 11	7
50–99	ⅢⅠ ⅢⅠ 1	11
100–149	ⅢⅠ 11	7
150–199	11	2
200–249		0
250–299	1	1

Why did you chose those groups? Why didn't you take 0 to 50 then 50 to 100 and so on? That would be much easier.

> Just stop and think for a minute. Look at **your** groups. Where would I count an hotel with 50 rooms? Would I count it twice?

> No, that can't be right. Of course you can't count something twice in two different groups. You were right – I can see why you chose those groups now.

▽ Try some for yourself

3. (a) Compile a grouped frequency table to show the price distribution in the hotels. Use the groups 40 to 44, 45 to 49, ...

(b) Compile a grouped frequency table for the same data using larger groups. This time use the groups 40 to 49, 50 to 59, ...

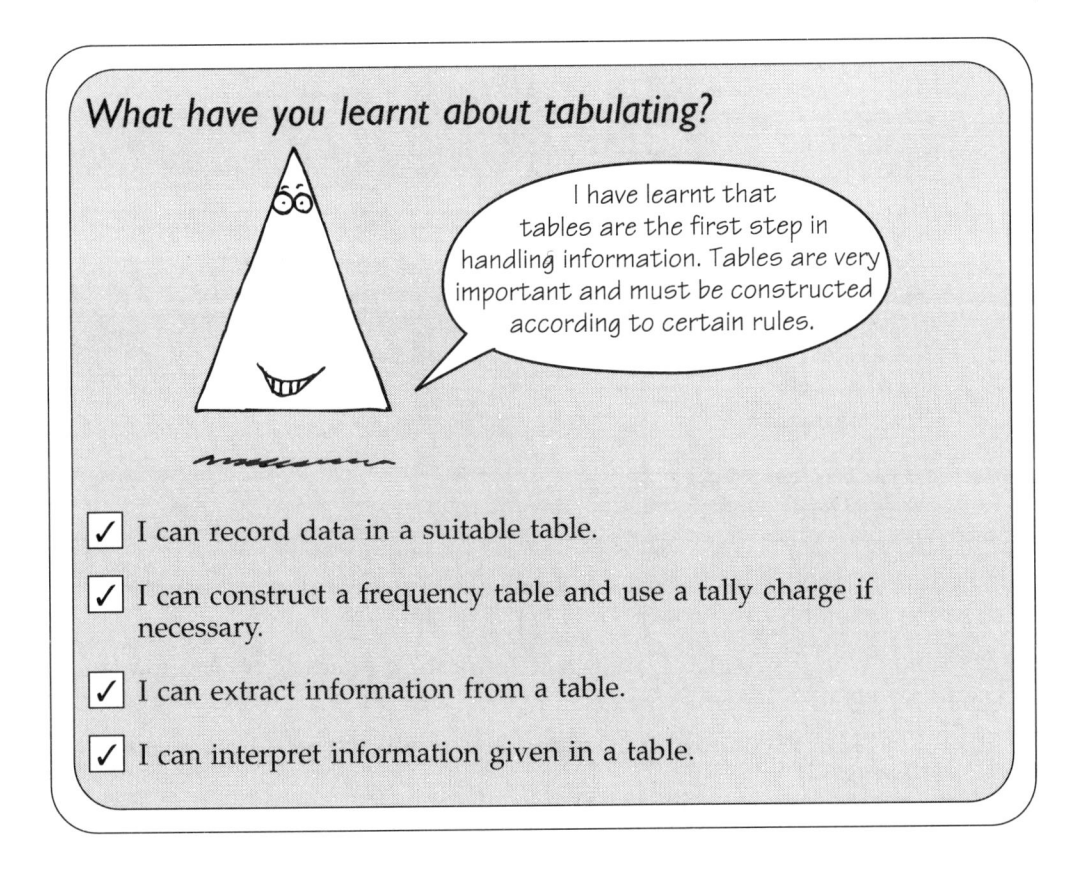

What have you learnt about tabulating?

> I have learnt that tables are the first step in handling information. Tables are very important and must be constructed according to certain rules.

- [✓] I can record data in a suitable table.
- [✓] I can construct a frequency table and use a tally charge if necessary.
- [✓] I can extract information from a table.
- [✓] I can interpret information given in a table.

Further Questions

1. Use the data in the table on p. 116 to construct four small tables, one for each category of hotel.

Each table should include details of:

(a) number of rooms

(b) price

(c) swimming pool.

2. Plymouth – Torbay – Exeter – London

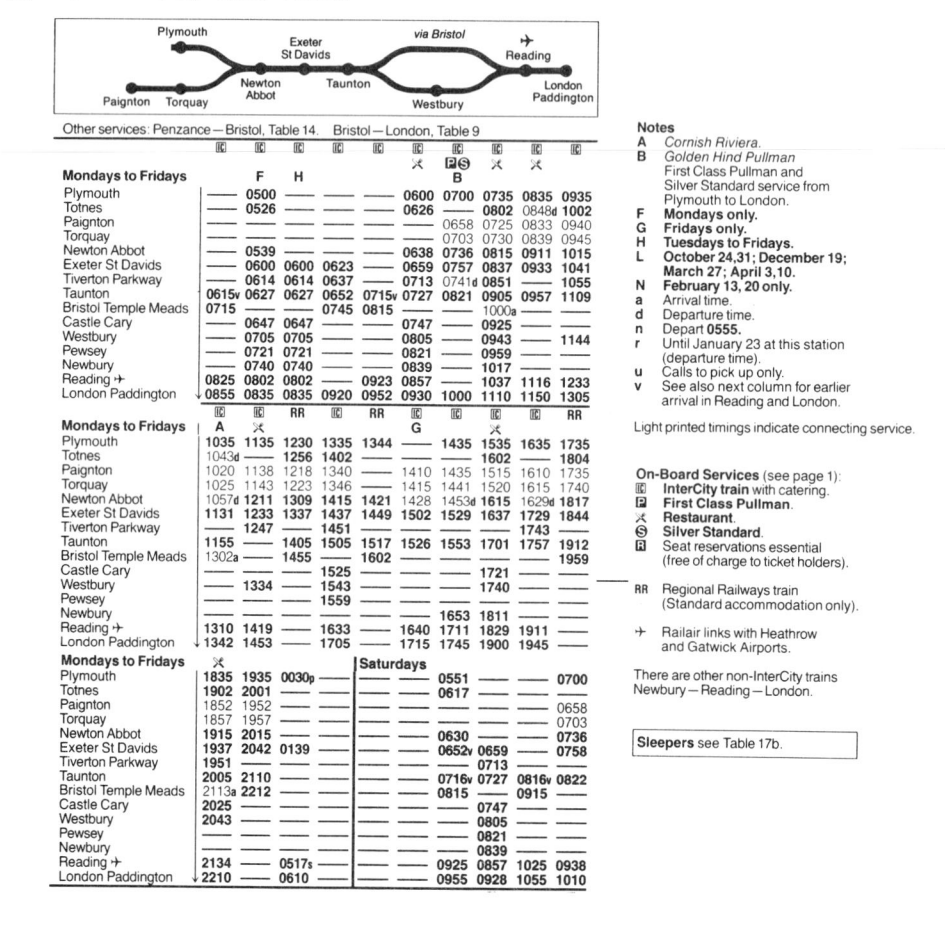

```
                    Plymouth                     via Bristol
                              Exeter                      ✈
                              St Davids                Reading
                        Newton      Taunton                 London
                        Abbot                               Paddington
            Paignton Torquay              Westbury
```

Other services: Penzance – Bristol, Table 14.   Bristol – London, Table 9.

Mondays to Fridays		F	H					PS B			
Plymouth	—	0500	—	—	—	0600	0700	0735	0835	0935
Totnes	—	0526	—	—	—	0626	—	0802	0848d	1002
Paignton	—	—	—	—	—	—	0658	0725	0833	0940
Torquay	—	—	—	—	—	—	0703	0730	0839	0945
Newton Abbot	—	0539	—	—	—	0638	0736	0815	0911	1015
Exeter St Davids	—	0600	0600	0623	—	0659	0757	0837	0933	1041
Tiverton Parkway	—	0614	0614	0637	—	0713	0741d	0851	—	1055
Taunton	0615v	0627	0627	0652	0715v	0727	0821	0905	0957	1109
Bristol Temple Meads	0715	—	—	0745	0815	—	—	1000a	—	—
Castle Cary	—	0647	0647	—	—	0747	—	0925	—	—
Westbury	—	0705	0705	—	—	0805	—	0943	—	1144
Pewsey	—	0721	0721	—	—	0821	—	0959	—	—
Newbury	—	—	—	—	—	0839	—	1017	—	—
Reading ✈	—	0825	0802	0802	—	0923	0857	—	1037	1116	1233
London Paddington	↓0855	0835	0835	0920	0952	0930	1000	1110	1150	1305

Mondays to Fridays	A	✕		RR		RR		G				RR
Plymouth	1035	1135	1230	1335	1344	—	1435	1535	1635	1735
Totnes	1043d	—	1256	1402	—	—	1602	—	1804
Paignton	1020	1138	1218	1340	—	1410	1435	1515	1610	1735
Torquay	1025	1143	1223	1346	—	1415	1441	1520	1615	1740
Newton Abbot	1057d	1211	1309	1415	1421	1428	1453d	1615	1629d	1817
Exeter St Davids	1131	1233	1337	1437	1449	1502	1529	1637	1729	1844
Tiverton Parkway	—	1247	—	1451	—	—	—	—	1743	—
Taunton	1155	—	1405	1505	1517	1526	1553	1701	1757	1912
Bristol Temple Meads	1302a	—	1455	—	1602	—	—	—	—	1959
Castle Cary	—	—	—	1525	—	—	—	1721	—	—
Westbury	—	1334	—	1543	—	—	—	1740	—	—
Pewsey	—	—	—	1559	—	—	—	—	—	—
Newbury	—	—	—	—	—	—	—	1653	1811	—
Reading ✈	1310	1419	—	1633	—	1640	1711	1829	1911	—
London Paddington	↓1342	1453	—	1705	—	1715	1745	1900	1945	—

Mondays to Fridays	✕				Saturdays					
Plymouth	1835	1935	0030p	—	—	0551	—	—	0700
Totnes	1902	2001	—	—	—	0617	—	—	—
Paignton	1852	1952	—	—	—	—	—	—	0658
Torquay	1857	1957	—	—	—	—	—	—	0703
Newton Abbot	1915	2015	—	—	—	0630	—	—	0736
Exeter St Davids	1937	2042	0139	—	—	0652v	0659	—	0758
Tiverton Parkway	1951	—	—	—	—	0713	—	—	—
Taunton	2005	2110	—	—	—	0716v	0727	0816v	0822
Bristol Temple Meads	2113a	2212	—	—	—	0815	—	0915	—
Castle Cary	2025	—	—	—	—	—	0747	—	—
Westbury	2043	—	—	—	—	—	0805	—	—
Pewsey	—	—	—	—	—	—	0821	—	—
Newbury	—	—	—	—	—	—	0839	—	—
Reading ✈	2134	—	0517s	—	—	0925	0857	1025	0938
London Paddington	↓2210	—	0610	—	—	0955	0928	1055	1010

Notes
- A Cornish Riviera.
- B Golden Hind Pullman First Class Pullman and Silver Standard service from Plymouth to London.
- F Mondays only.
- G Fridays only.
- H Tuesdays to Fridays.
- L October 24, 31; December 19; March 27; April 3, 10.
- N February 13, 20 only.
- a Arrival time.
- d Departure time.
- n Depart 0555.
- r Until January 23 at this station (departure time).
- u Calls to pick up only.
- v See also next column for earlier arrival in Reading and London.

Light printed timings indicate connecting service.

On-Board Services (see page 1):
- IC InterCity train with catering.
- P First Class Pullman
- ✕ Restaurant.
- S Silver Standard.
- R Seat reservations essential (free of charge to ticket holders).
- RR Regional Railways train (Standard accommodation only).
- ✈ Railair links with Heathrow and Gatwick Airports.

There are other non-InterCity trains Newbury – Reading – London.

Sleepers see Table 17b.

The rail timetable shows services from Devon to London Paddington. Use the timetables to produce smaller timetables to show:

(a) Trains from Exeter St. Davids via Taunton to Reading on Mondays to Fridays.

(b) Trains from Plymouth to London Paddington where a restaurant car service is available.

Solutions

1.

Number of rooms	Tally	Frequency
1	111	3
2	1111	5
3	1	1
4		0
5	1	1
	Total	10

2. (a) 28

   (b) 3.57%

   (c) 64.29%

3. (a)

Number of rooms	Tally	Frequency
40–44	11	2
45–49	1111 1111 1	11
50–54	1111 1	6
55–59	1111	4
60–64	11	2
65–69		0
70–74	11	2
75–79	1	1
	Total	28

(b)

Number of rooms	Tally	Frequency
40–49	1111 1111 111	13
50–59	1111 1111	10
60–69	11	2
70–79	111	3
	Total	28

121

9 Illustrating Information

I am finding that a lot of my lecturers are asking me to use graphs to illustrate my assignments. I know how to draw bar charts and pie charts but I'm not so confident about pictograms and more complicated bar charts.

Would you like me to go over all the forms of pictorial representation that you are expected to be able to use?

Yes, that would be useful. Could you give me some hints on which graphs to use for different types of information? Usually I use one of every type just to be sure!

Each type of graph gives a different kind of information. It is important to know how to match your graph to your information. This knowledge will be useful when you are asked to comment on your own graphs or ones you are given.

122

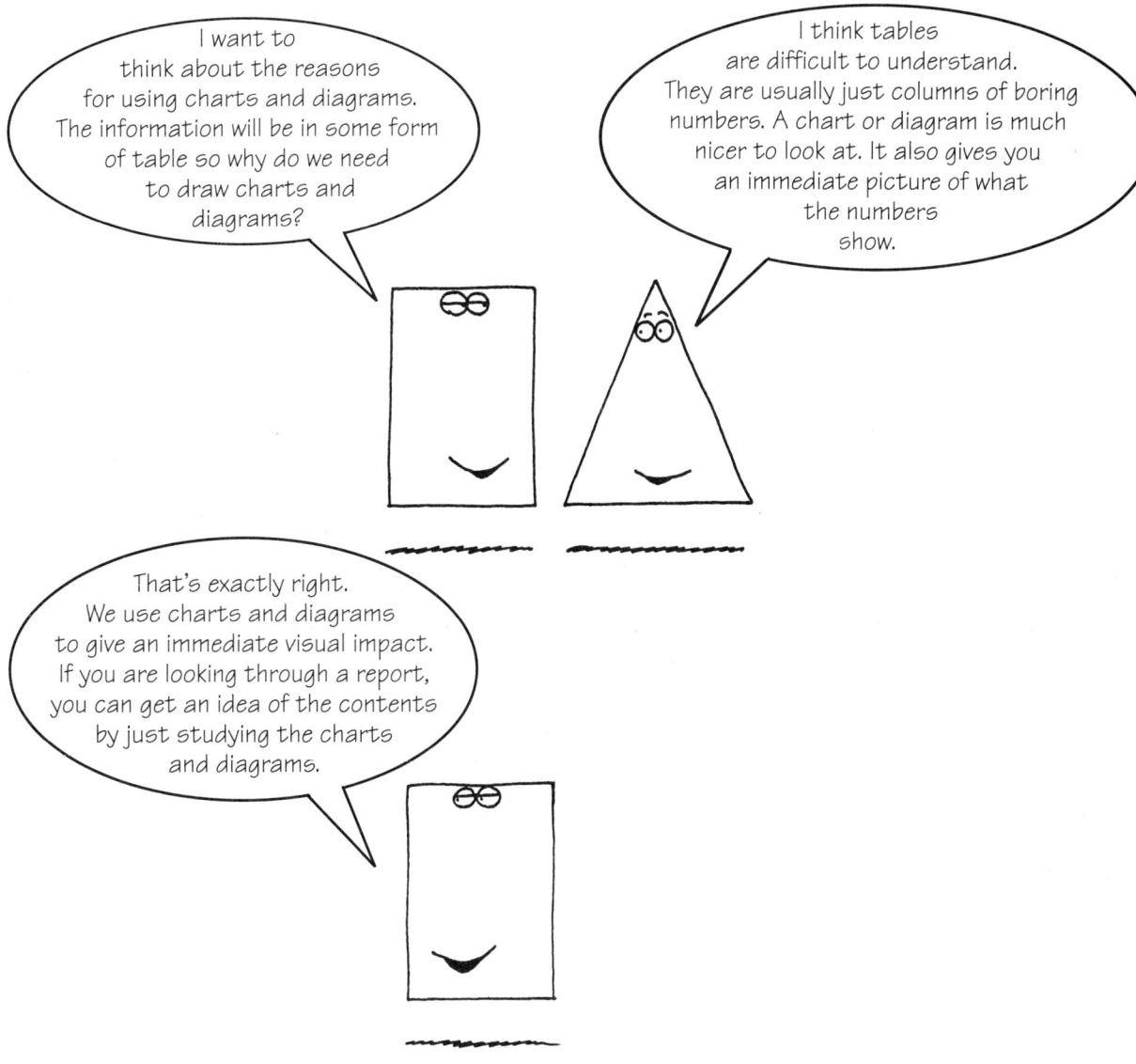

Using charts and diagrams

> I want to think about the reasons for using charts and diagrams. The information will be in some form of table so why do we need to draw charts and diagrams?

> I think tables are difficult to understand. They are usually just columns of boring numbers. A chart or diagram is much nicer to look at. It also gives you an immediate picture of what the numbers show.

> That's exactly right. We use charts and diagrams to give an immediate visual impact. If you are looking through a report, you can get an idea of the contents by just studying the charts and diagrams.

Charts and diagrams:

- make it easier for us to grasp the size of the figures involved.
- make it easier for us to see how sets of figures are related.
- make it easier for us to make comparisons.

Guidelines for the production of charts and diagrams

- All charts and diagrams must be titled and the source of the information indicated.
- All charts and diagrams must be clearly labelled.
- All charts and diagrams must be as clear and simple as possible.
- All charts and diagrams must have obvious meanings.
- All charts and diagrams must contain all the relevant information.

Pictograms

The first diagram I want to look at is really a picture. I think we should look at some pictograms and find out what they are used for.

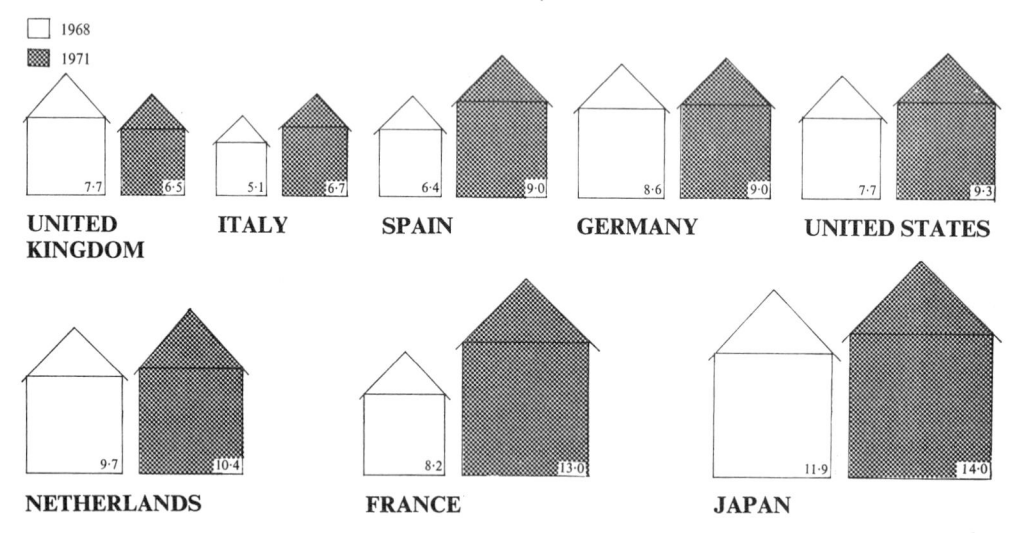

DWELLINGS COMPLETED PER 1000 INHABITANTS

☐ 1968
▩ 1971

| UNITED KINGDOM | ITALY | SPAIN | GERMANY | UNITED STATES |
| 7·7 6·5 | 5·1 6·7 | 6·4 9·0 | 8·6 9·0 | 7·7 9·3 |

| NETHERLANDS | FRANCE | JAPAN |
| 9·7 10·4 | 8·2 13·0 | 11·9 14·0 |

Source: O.E.C.D.

This pictogram is illustrating information about house building. The symbol used is very simple and easy to draw. The size of the symbol is varied to match the figure it represents.

A weather chart is a form of pictogram. Symbols are used to indicate weather conditions. A key to their meaning is given so the reader can understand the information.

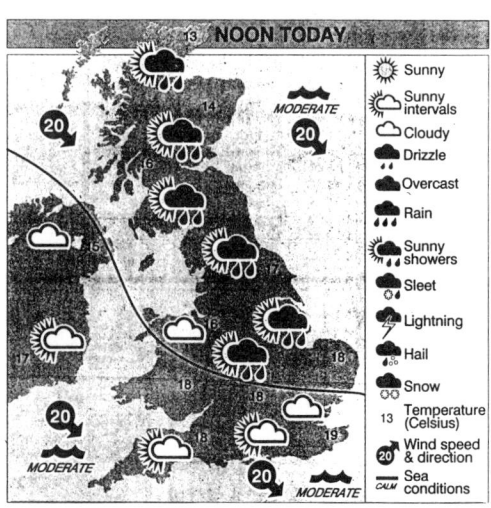

NOON TODAY

Key:
- Sunny
- Sunny intervals
- Cloudy
- Drizzle
- Overcast
- Rain
- Sunny showers
- Sleet
- Lightning
- Hail
- Snow
- 13 Temperature (Celsius)
- 20 Wind speed & direction
- Sea conditions

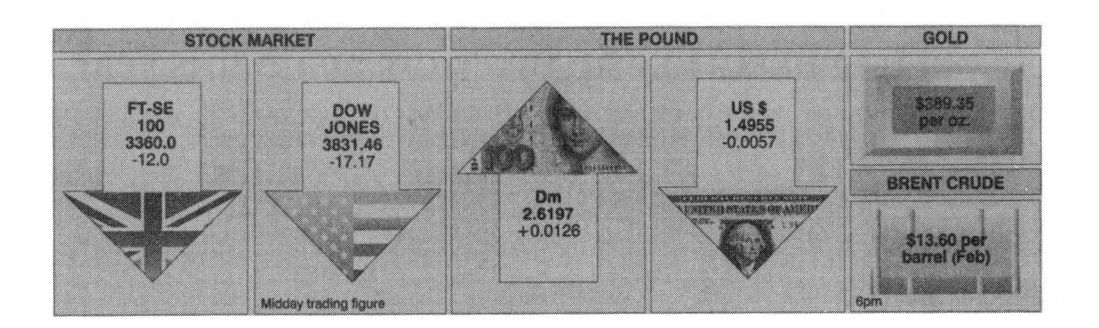

This pictogram is providing information about closing prices on the London Stock Exchange.
National flags are used to indicate currencies.
An ingot is used to symbolise gold.
A barrel is used to symbolise oil.

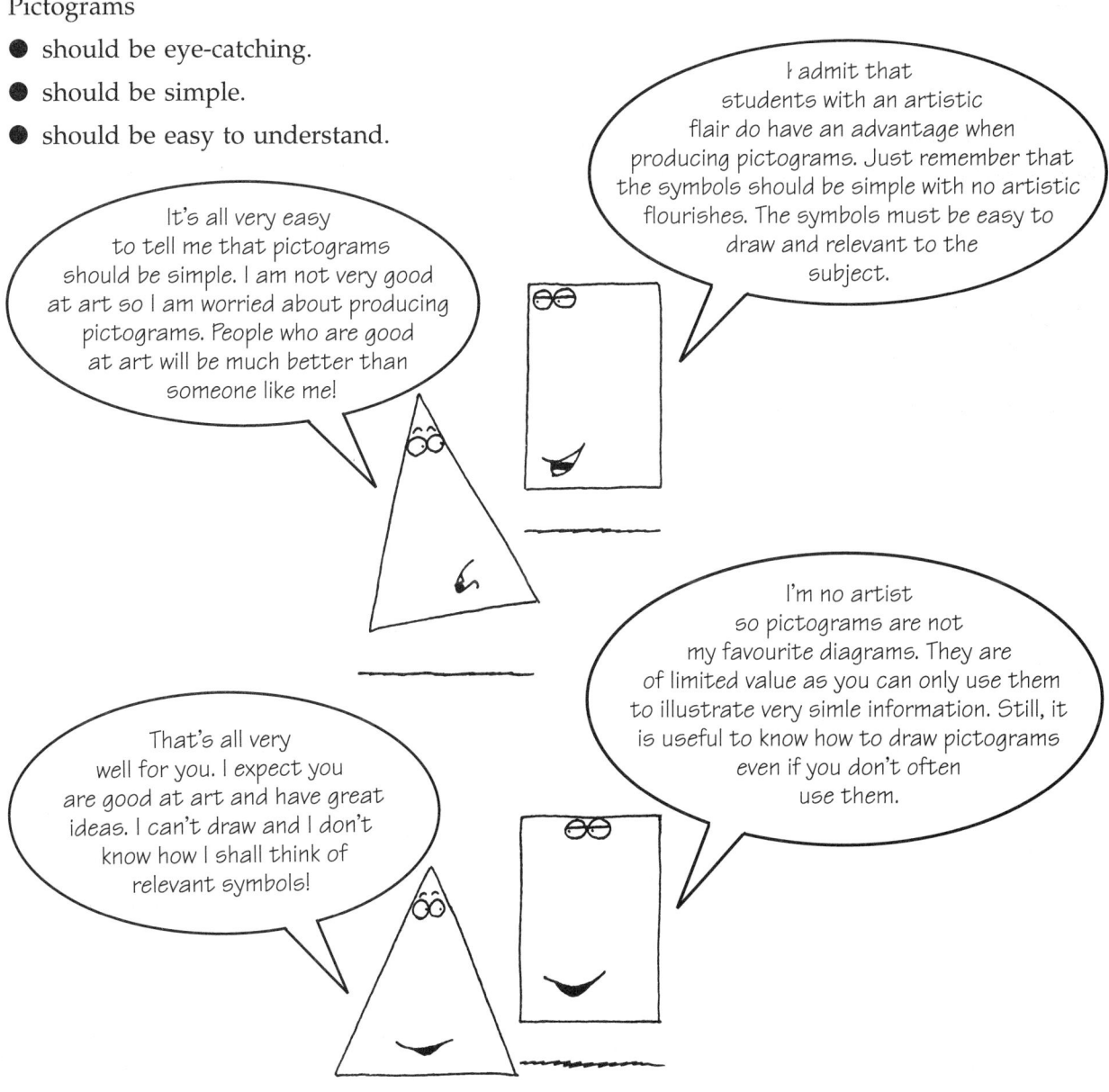

Pictograms

- should be eye-catching.
- should be simple.
- should be easy to understand.

It's all very easy to tell me that pictograms should be simple. I am not very good at art so I am worried about producing pictograms. People who are good at art will be much better than someone like me!

I admit that students with an artistic flair do have an advantage when producing pictograms. Just remember that the symbols should be simple with no artistic flourishes. The symbols must be easy to draw and relevant to the subject.

That's all very well for you. I expect you are good at art and have great ideas. I can't draw and I don't know how I shall think of relevant symbols!

I'm no artist so pictograms are not my favourite diagrams. They are of limited value as you can only use them to illustrate very simle information. Still, it is useful to know how to draw pictograms even if you don't often use them.

125

Let's look at a simple pictogram giving information about the use of light bulbs in a three-bedroomed house.

Electric light bulbs used in a 3-bedroomed house

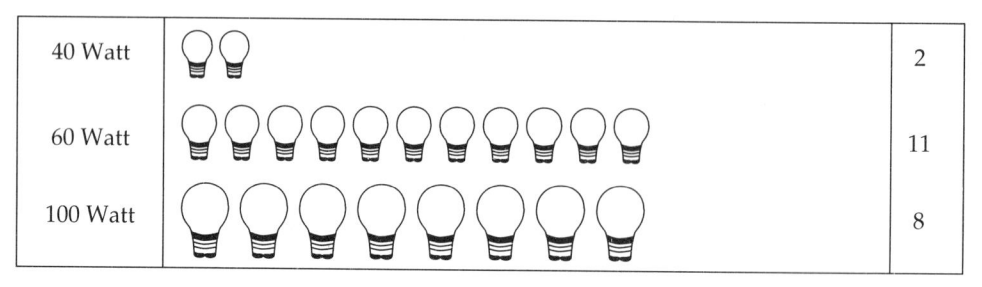

40 Watt	🔆🔆	2
60 Watt	💡💡💡💡💡💡💡💡💡💡💡	11
100 Watt	💡💡💡💡💡💡💡💡	8

The symbols used are very simple and very relevant. Different sizes have been used to indicate the different sizes of the bulbs.

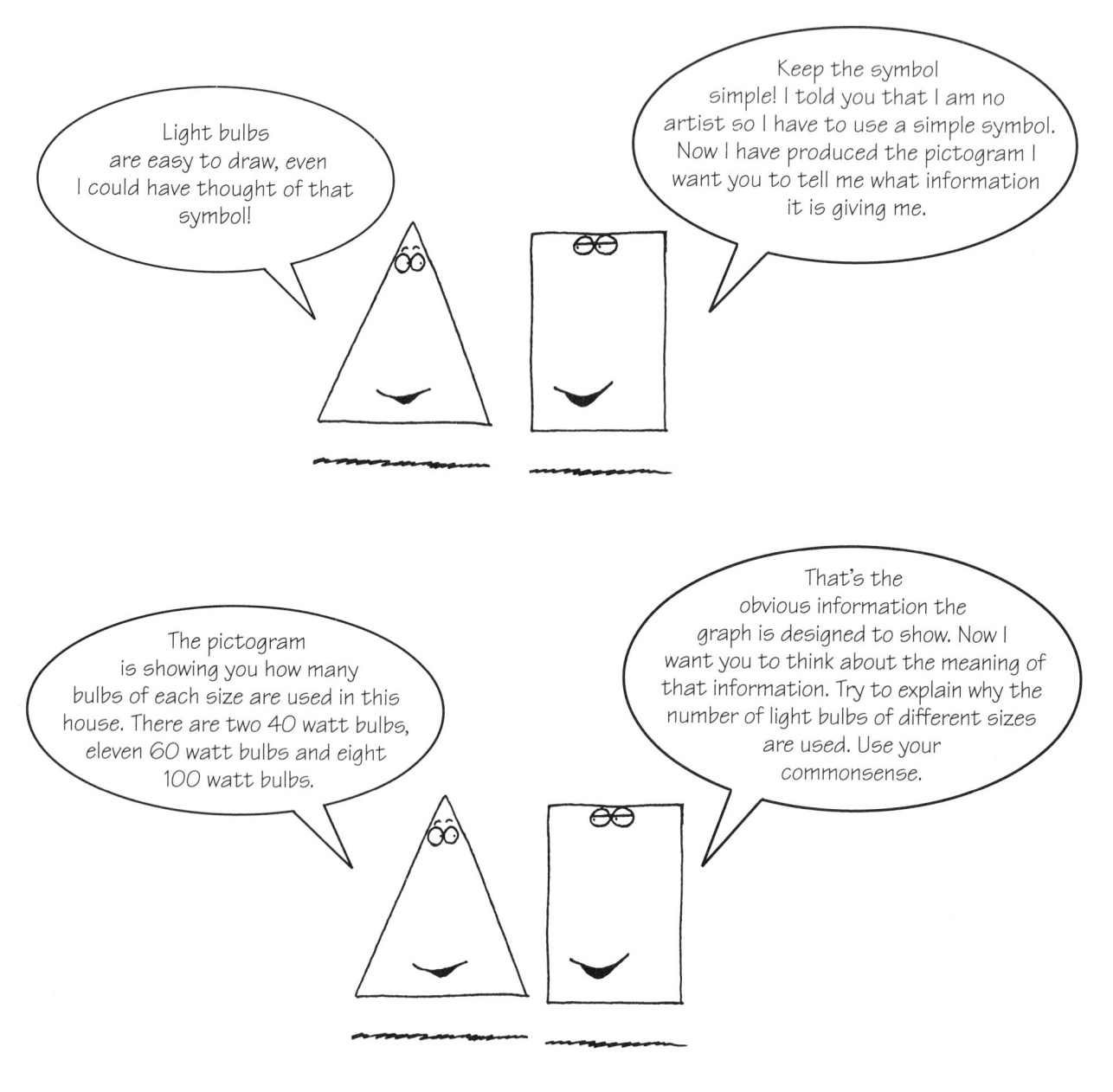

> Light bulbs are easy to draw, even I could have thought of that symbol!

> Keep the symbol simple! I told you that I am no artist so I have to use a simple symbol. Now I have produced the pictogram I want you to tell me what information it is giving me.

> The pictogram is showing you how many bulbs of each size are used in this house. There are two 40 watt bulbs, eleven 60 watt bulbs and eight 100 watt bulbs.

> That's the obvious information the graph is designed to show. Now I want you to think about the meaning of that information. Try to explain why the number of light bulbs of different sizes are used. Use your commonsense.

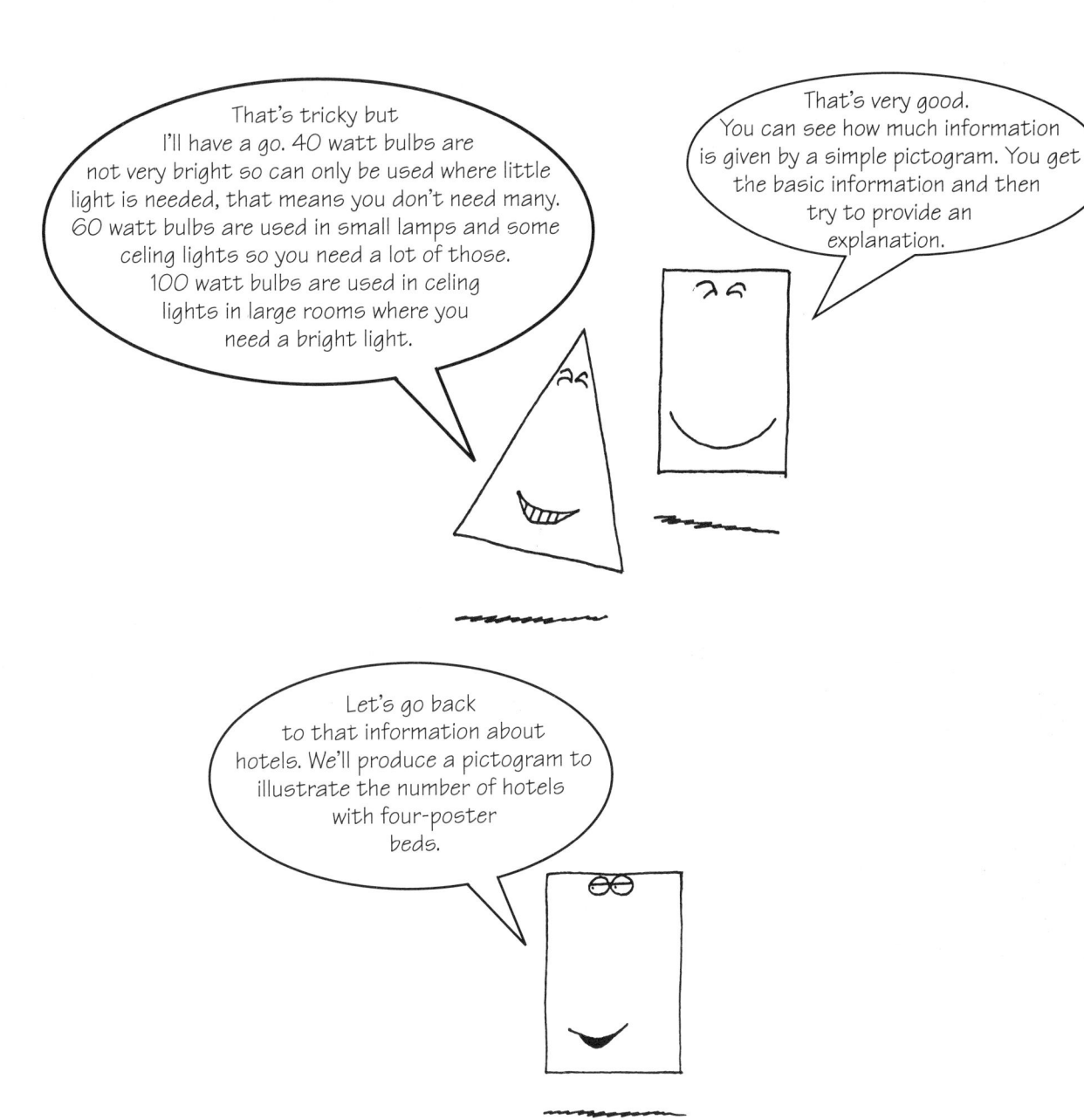

That's tricky but I'll have a go. 40 watt bulbs are not very bright so can only be used where little light is needed, that means you don't need many. 60 watt bulbs are used in small lamps and some celing lights so you need a lot of those. 100 watt bulbs are used in celing lights in large rooms where you need a bright light.

That's very good. You can see how much information is given by a simple pictogram. You get the basic information and then try to provide an explanation.

Let's go back to that information about hotels. We'll produce a pictogram to illustrate the number of hotels with four-poster beds.

Number of four-poster beds	Number of hotels
1	3
2	5
3	1
4	0
5	1
Total	10

We need to design a symbol for an hotel. Remember it must be simple because we have to draw a total of 12 identical ones.

⌂ = 1 hotel

Pictogram showing the number of hotels with four-poster beds

Number of four-poster beds	
1	⌂ ⌂ ⌂
2	⌂ ⌂ ⌂ ⌂ ⌂
3	⌂
4	
5	⌂

At a glance we can see that 5 hotels have 2 four-poster beds, 3 have one each, one has 3 and another has 5.

As these hotels are located in the West Midlands, which includes 'Shakespeare Country', it is not so surprising to find so many hotels with four-poster beds. Many hotels in this area are located in historic buildings so the beds were probably there before the buildings became hotels. Four-poster beds are popular, particularly with honeymoon couples so hotels are keen to offer this facility.

▽ Try some for yourself

1. Using the hotels data, produce a pictogram to show the number of hotels in each category. Remember to chose a simple symbol – maybe an initial letter?

Category	Total
C	4
G	2
H	16
P	6

Comment on the information displayed in the pictogram.

2. The table gives the average daily hours of sunshine in Tunisia during the summer season. Use a pictogram to illustrate this information. Interpret the information fro the pictogram.

Month	April	May	June	July	Aug	Sept	Oct
Hours of sunshine	8	10	12	12	11	9	7

3. A group of GNVQ students recorded their shoes sizes.

Size	4	5	6	7
Number	1	3	2	1

Use a pictogram to illustrate this information.

4. A GNVQ tutor group was made up of 10 male students and 8 female students.

Use a pictogram to illustrate this information.

Pie charts

Now you know how to draw pictograms it's time to move on to pie charts. These are very useful and quite easy to draw. No artistic talent required!

Pie charts are used to show proportions – how big a slice of the pie each item represents.

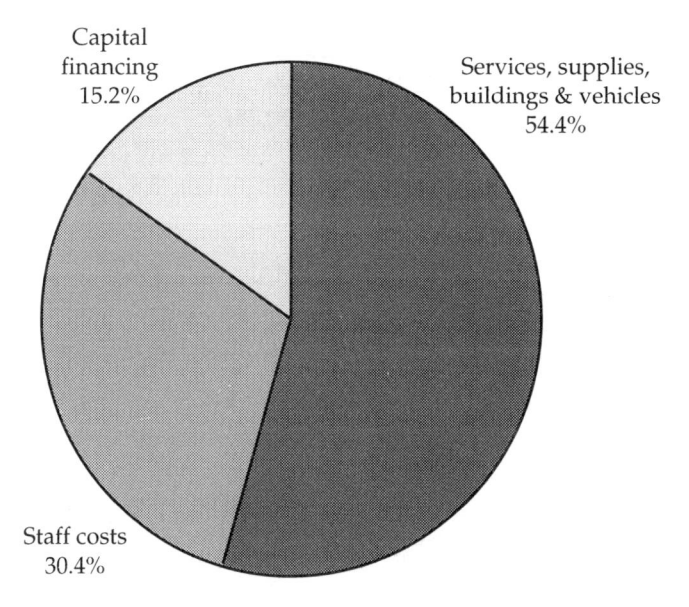

Council tax
How your money was spent

Capital financing 15.2%

Services, supplies, buildings & vehicles 54.4%

Staff costs 30.4%

Here we can see that over half of the council tax was used for services, supplies, buildings and vehicles.

Staff costs accounted for just over one quarter and the smallest slice was used for capital financing.

Pie charts:

- are a dramatic and visually appealing way to illustrate information.
- make it easy to grasp the significance of the information presented.
- show the relationship of the parts to the whole.

Pie charts are very useful but have their limitations.

- Pie charts require special equipment. You must have, and be able to use, compasses and a protractor.
  It is useful to have a circular protractor or even a special pie-chart template.
- Pie charts are not particularly accurate as the angles at the centre must total 360° exactly.
  Sometimes the figures have to be rounded or otherwise adjusted to obtain that total.
- Pie charts can become messy if too much information is included. As a general rule a maximum of six divisions is the limit. Too many divisions and colours can be very messy.

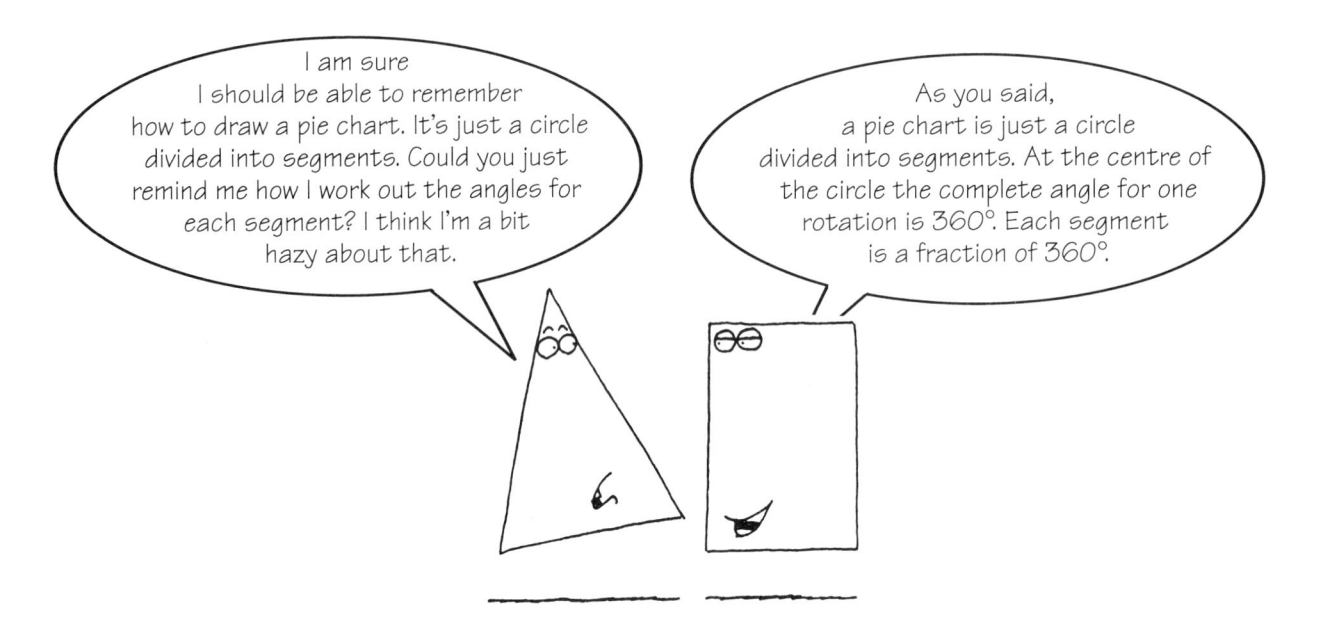

> I am sure I should be able to remember how to draw a pie chart. It's just a circle divided into segments. Could you just remind me how I work out the angles for each segment? I think I'm a bit hazy about that.

> As you said, a pie chart is just a circle divided into segments. At the centre of the circle the complete angle for one rotation is 360°. Each segment is a fraction of 360°.

Let's look at a simple example.

We are going to calculate the angles for a piechart to illustrate this data.

Colour of eyes in a GNVQ tutor group

Blue eyes	11
Brown eyes	8
Green eyes	3
Grey eyes	2
Total	24

Blue eyes = $\frac{11}{24}$ of the total so the angle is $\frac{11}{24} \times 360°$ = 165°

Brown eyes = $\frac{8}{24}$ of the total so the angle is $\frac{8}{24} \times 360°$ = 120°

Green eyes = $\frac{3}{24}$ of the total so the angle is $\frac{3}{24} \times 360°$ = 45°

Grey eyes = $\frac{2}{24}$ of the total so the angle is $\frac{2}{24} \times 360°$ = 30°

Total = 360°

Pie chart showing the colour of eyes of students in a GNVQ tutor group

We can see that the largest slice of the pie is representing blue eyes. Green and grey eyes occur far less often so have small slices. Brown eyes are quite common but are not as common as blue eyes.

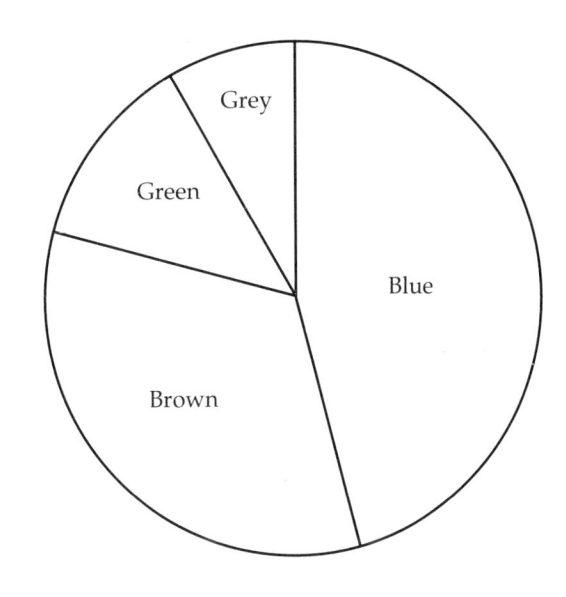

In this example, all the angles were whole numbers but this does not often happen. Usually the angles calculated are to a number of decimal places so you need to be able to round up or down to the nearest whole number.

Here is some data about drug usage.

All the figures are in percentages so the total will be 100.

Age when first tried drugs

Under 12 years old	4%
12–14 years	27%
14–16 years	31%
16–20 years	25%
21 or more years	13%

$$4\% = \frac{4}{100} \times 360° = 14.4° \qquad\qquad 14°$$

$$27\% = \frac{27}{100} \times 360° = 97.2° \qquad\qquad 97°$$

$$31\% = \frac{31}{100} \times 360° = 111.6° \qquad\qquad 112°$$

$$25\% = \frac{25}{100} \times 360° = 90° \qquad\qquad 90°$$

$$13\% = \frac{13}{100} \times 360° = 46.8° \qquad\qquad 47°$$

To the nearest whole number

Pie chart showing the ages at which people were introduced to drugs

Here we see that the ages from 12 to 20 years are the most susceptible. The 14–16 year period is the most likely time when teenages are trying to conform to peer group pressure. Few people used drugs before the age of 12 and only 13% started after the age of 21. This indicates that the teenage years are when young people are most likely to experiment with drugs.

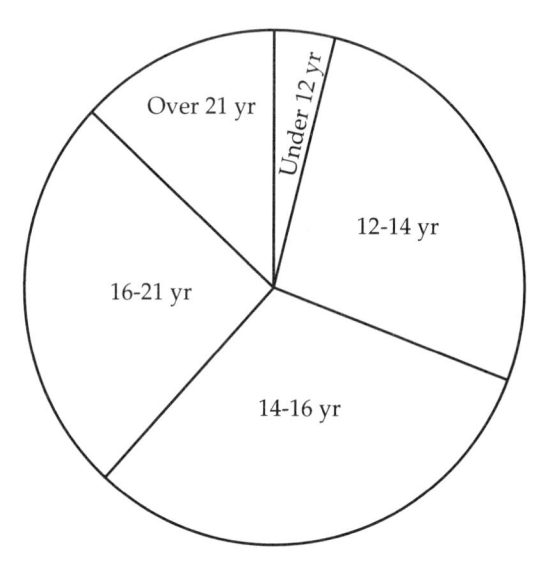

▽ Try some for yourself

5. Draw a pie chart to illustrate the following data. Comment on the information given in the diagram.

Number of Forte hotels in each category

Category	Total	Angle calculated	Angle to nearest whole number
C	4		
G	2		
H	16		
P	6		
Totals	28		360°

132

Line graphs

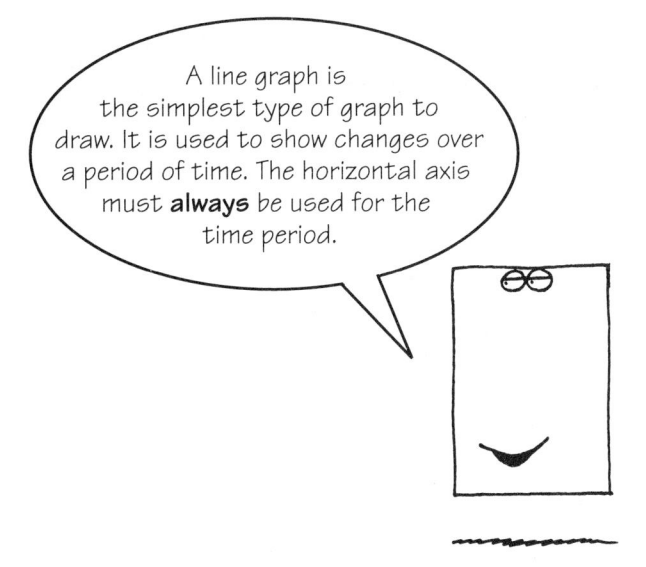

*A line graph is the simplest type of graph to draw. It is used to show changes over a period of time. The horizontal axis must **always** be used for the time period.*

Purchasing power of a 1951 pound

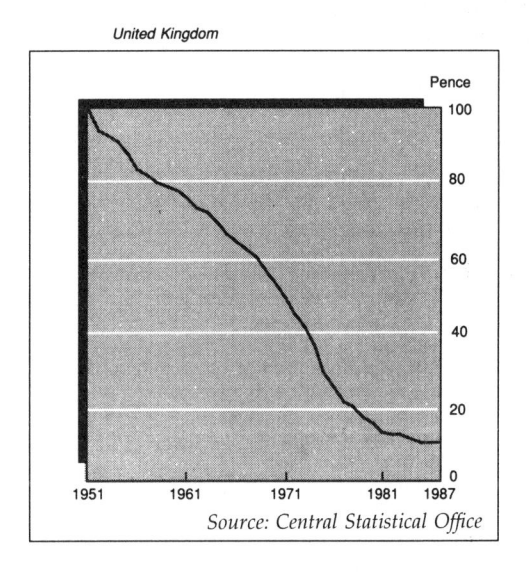

This line graph shows the decline in the purchasing power of the pound at 1951 values. The decline was steadily downwards but began to level out in 1987.

Source: Central Statistical Office

Private medical insurance (UK)

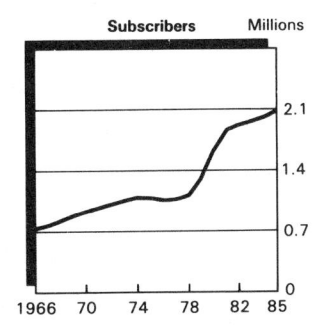

This line graph shows the rise in the number of people subscribing to private health insurance. The rise was slow between 1966 and 1974. There was a decline until 1978 when the numbers began to rise more rapidly. This rapid rise continued until 1982 when the rise slowed but the figures continued to increase.

Source: Social Trends 1987

I understand how to draw a line graph. It's quite a simple one to construct. I am not sure I could get all that information from just looking at the graph. Perhaps it comes with practice!.

Interpreting a graph is largely a matter of commonsense. You look at how the graph rises and falls and interpret those fluctuations. Now look carefully at this graph and tell me what you can see.

Graph showing maths test scores for a GNVQ student

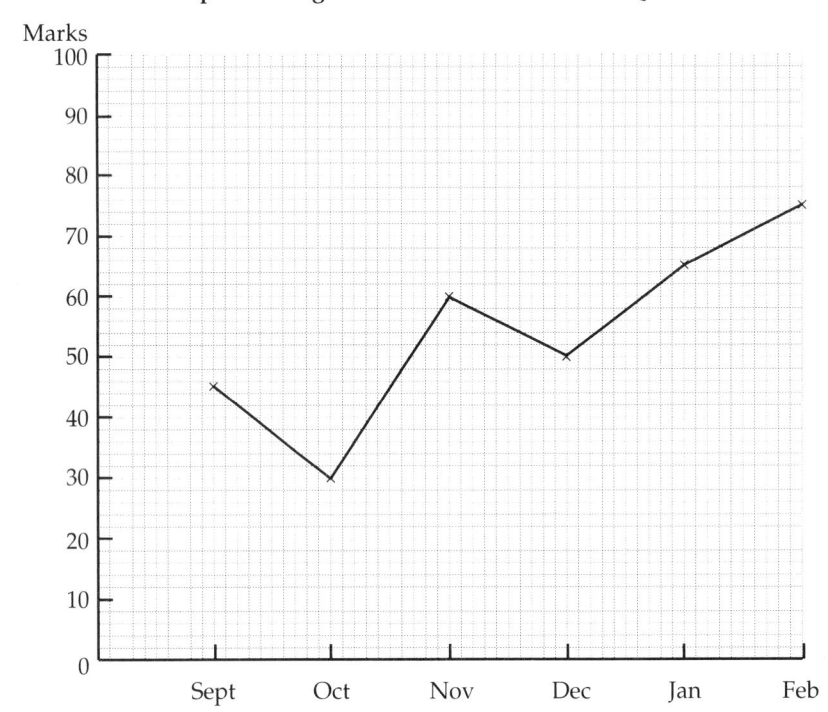

That's quite easy. The student's score was 45 in September and dropped to 30 in October. In November it rose to 60 but dropped down to 50 in December. In January it went up to 65 and climbed to 75 in February.

You have identified the ups and downs of the scores. Now I want you to try to explain why the scores go up and down as they do. There's no technical explanation – just use your commonsense.

134

> I'll have a go.
> When the student began the course in September he was not working very hard – probably everything was new to him and he took some time to settle in. His tutor may have talked to him and advised him to get some extra help from a maths workshop. He began to improve but slacked off in December – too many parties. After Christmas he really started to work and his scores got steadily better.

> That is an excellent interpretation of the graph. Remember that there is no single correct explanation because we don't know all the facts. We can only use our commonsense.

Try some for yourself

6. A clothing shop produces its trading figures for the three months from December to February. These figures show weekly sales over the period.

Month	December				January				February			
Week beginning	7	16	21	28	5	12	19	26	2	9	16	23
Sales in thousands of pounds	17	20	22	29	31	30	11	9	8	6	7	5

(a) Draw a line graph to illustrate this data.

(b) Interpret the graph.

7. (a) Draw a graph to illustrate the data on hours of sunshine in Tunisia.

Month	April	May	June	July	Aug	Sept	Oct
Hours of sunshine	8	10	12	12	11	9	7

(b) Interpret the graph.
Use the information to advise holiday-makers thinking of going to Tunisia.

135

Bar charts

Bar charts are the most popular method of pictorial representation. There are several obvious reasons for their popularity.

- Bar charts are easy to draw.
- Bar charts are reasonably accurate.
- Values can be read off the vertical (or horizontal) scale.
- Bar charts are easy to understand.

Bar charts are used to show comparative sizes. The height of the bar indicates the number of items represented.

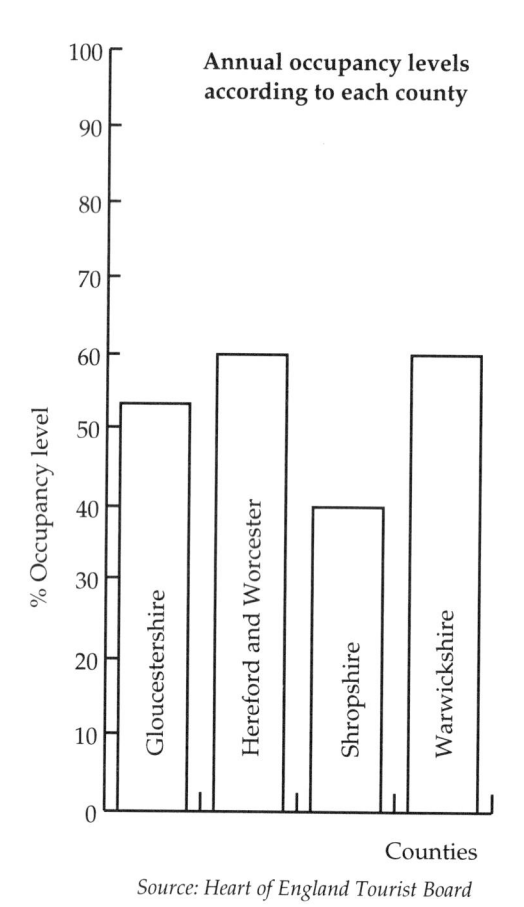

Annual occupancy levels according to each county

Source: Heart of England Tourist Board

This bar chart shows that the highest occupancy levels were in the counties of Hereford and Worcester and Warwickshire, the countries closely associated with Shakespeare. The lowest occupancy level was in Shropshire – a less popular tourist destination.

If appropriate a bar chart may be drawn horizontally.

Composition of total net expenditure: by sector of education 1986–8

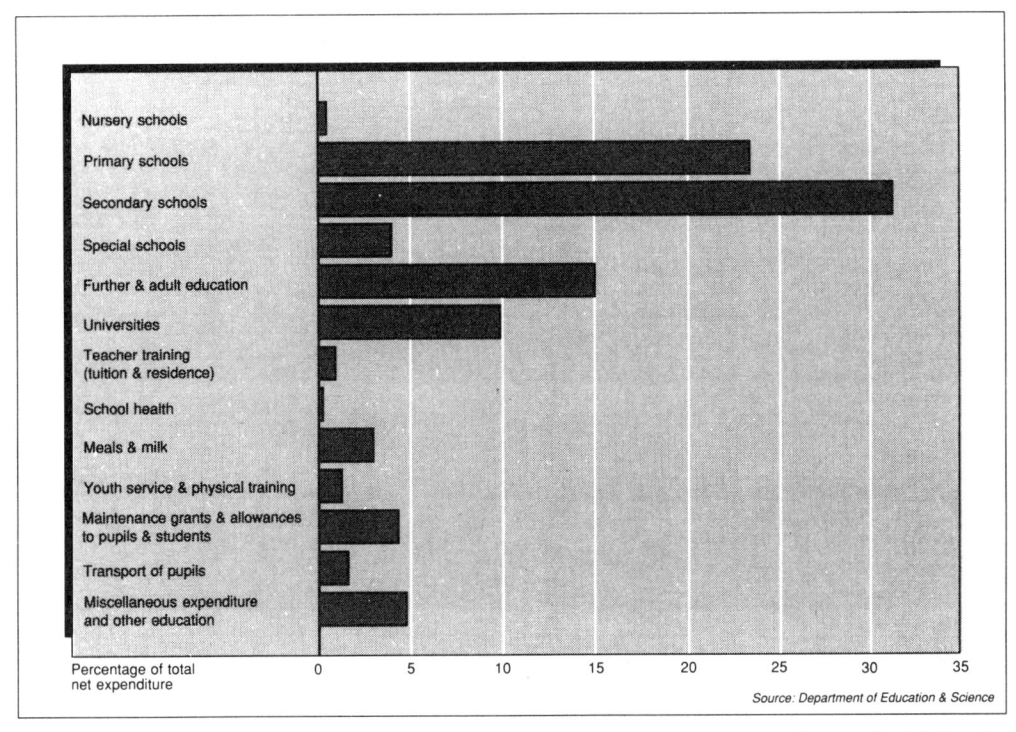

Source: Social Trends

This horizontal bar chart shows clearly that secondary schools received the largest percentage of the education budget. Primary schools received approximately 8% less than secondary schools. Nursery schools received less than 1% of the total budget.

Bar charts can be used to show negative values.

This bar chart shows that during the period from June 1983 to June 1988 approximately 150 thousand men became unemployed.

Negative employment means unemployment.

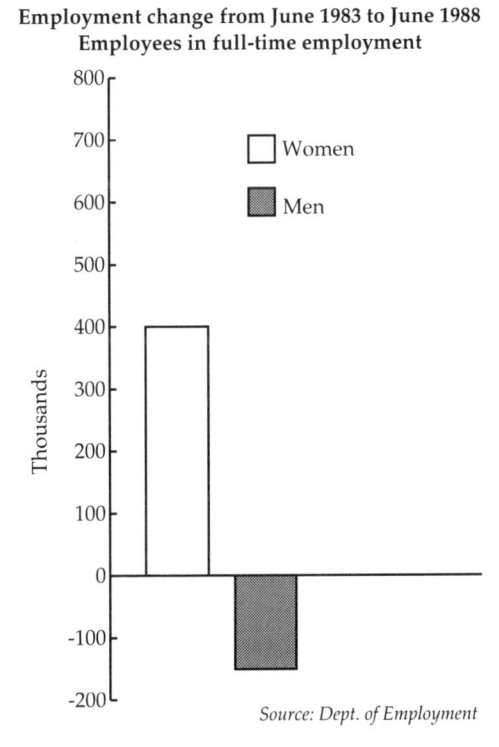

> Bar charts are easy to draw. I am sure I shall have no trouble with them. I need to check one or two points first just to make sure I'm right.

> Bar charts are easy to construct. All you need is a straight edge and a **sharp** pencil. There are some points to remember when drawing a bar chart.

- A bar chart needs vertical and horizontal axes.
- The vertical axis must be clearly labelled.
- The scale of values is marked on the vertical axis.
- The bars must be of equal width and evenly spaced.
- The heights of the bars are determined by the number of each item.

> I was going to ask about spaces between the bars. I'm sure I have seen bar charts with two or three bars next to each other.

> There are bar charts where bars are next to each other, and some where the bars are stacked on top of each other. Let's just concentrate on simple bar charts first.

Let's draw a bar chart to illustrate the data on eye colour.

Colour	Number of students
Blue	11
Brown	8
Green	3
Grey	2

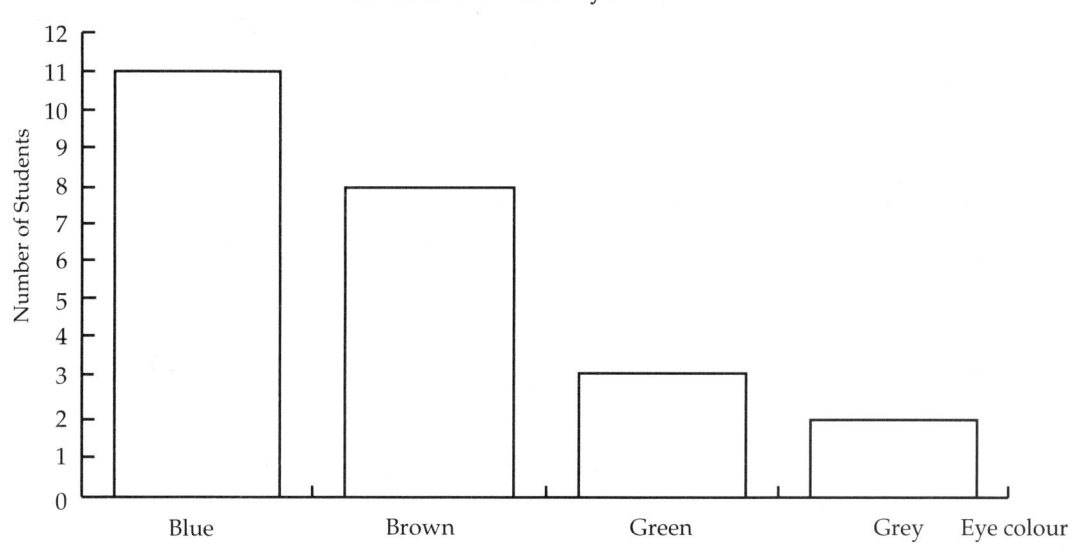

Bar chart of colour of eyes

Can you interpret this bar chart? Remember, a bar chart is used to illustrate comparisons.

Yes, this is easy! The most common eye colour is blue, with brown only 3 behind. Green and grey are much less common.

Let's draw a bar chart to illustrate the electric light bulb data.

Size	Number
40 watt	2
60 watt	11
100 watt	8

Notice that each bar is labelled below the horizontal axis.

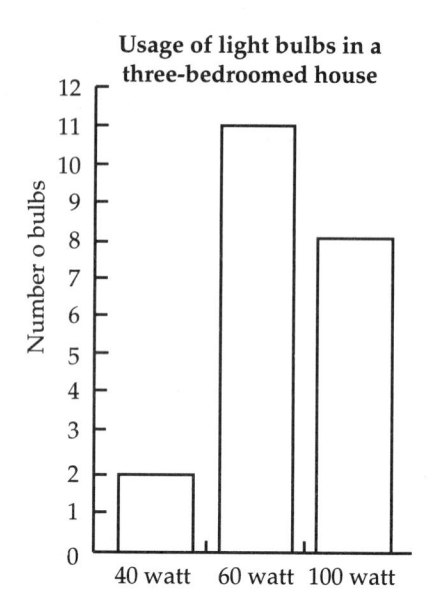

Usage of light bulbs in a three-bedroomed house

We can see very clearly that this household uses more 60 watt bulbs than any other type. The few 40 watt bulbs are obvious because of the relative heights of the bars.

▽ Try some for yourself

8. Number of hotels in each category

Category	Total
C	4
G	2
H	16
P	6

Draw a bar chart to illustrate this information.

9. **Data on drug usage**

Draw a bar chart to illustrate this information. Comment on the information displayed in your chart.

First drug ever taken

Cannabis	59%
Speed	12%
Tranquillisers	6%
Barbiturates	6%
LSD	5%
Solvents	5%
Heroin	3%
Other opiates	3%
Magic mushrooms	1%

Compound (or multiple) bar charts

Sometimes it is necessary to make direct comparisons between two or more pieces of information.

We may need to compare the same type of information from year to year.

We may need to compare several items within one year.

To illustrate such data we use a compound bar chart where bars are placed side by side.

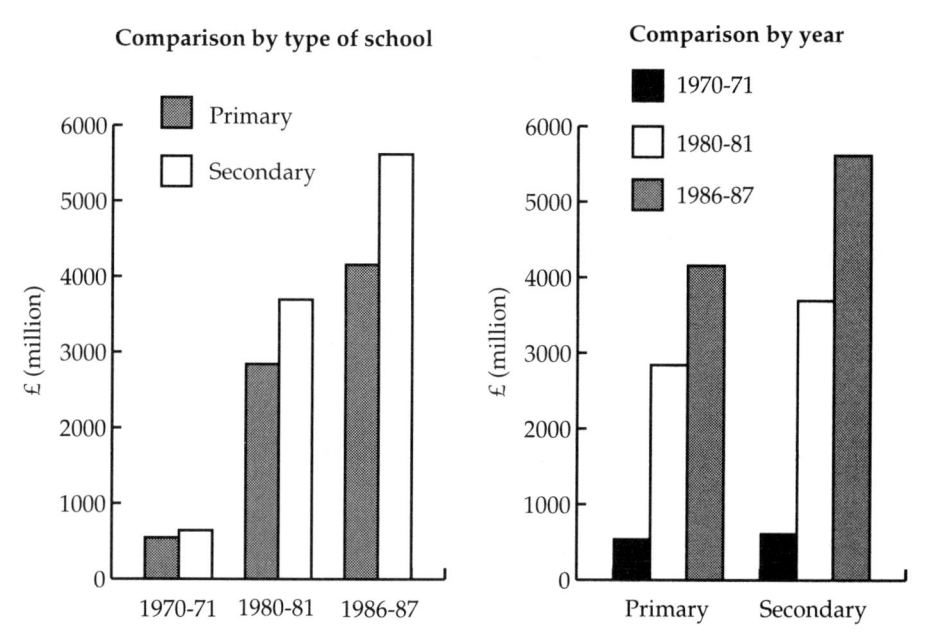

These compound bar charts illustrate the following data.

Government expenditure on education in the United Kingdom (£ millions)

	1970–1	1980–1	1986–7
Primary schools	546	2840	4157
Secondary schools	619	3695	5620

Source: Social Trends 1987

That's very clever! You have compared expenditure on primary and secondary schools for each year, then you have compared the annual expenditure for each type of school. That's really good.

Usually, if you have information that is best presented by a compound bar chart, you will find that you do have a choice about how to present it. There are always two ways.

141

▽ Try some for yourself

10. Here is a table of data relating to the initial suppliers of drugs.

Drug	Supplier		
	Friend	Family	Dealer
Heroin	64%	8%	28%
Cannabis	78%	3%	19%
Speed	84%	6%	10%

Draw compound bar charts to illustrate this data.

One set of charts should show comparison by drug.

One set of charts should show comparison by supplier.

Component bar charts

You mentioned a third type of bar chart. You said there is one where the bars are stacked on top of each other. What's that called?

That's called a component bar chart. Each bar is made up of a number of components. Instead of putting them side by side they are stacked into one bar.

Government expenditure on education in the United Kingdom (£ millions)

	1970–1	1980–1	1986–7	Totals
Primary schools	546	2840	4157	7543
Secondary schools	619	3695	5620	9934
Totals	1165	6535	9777	17 477

Source: Social Trends 1987

To draw a component bar chart we must first total the columns or rows to give us an idea of the scale we will need.

Component bar chart to illustrate government expenditure on education

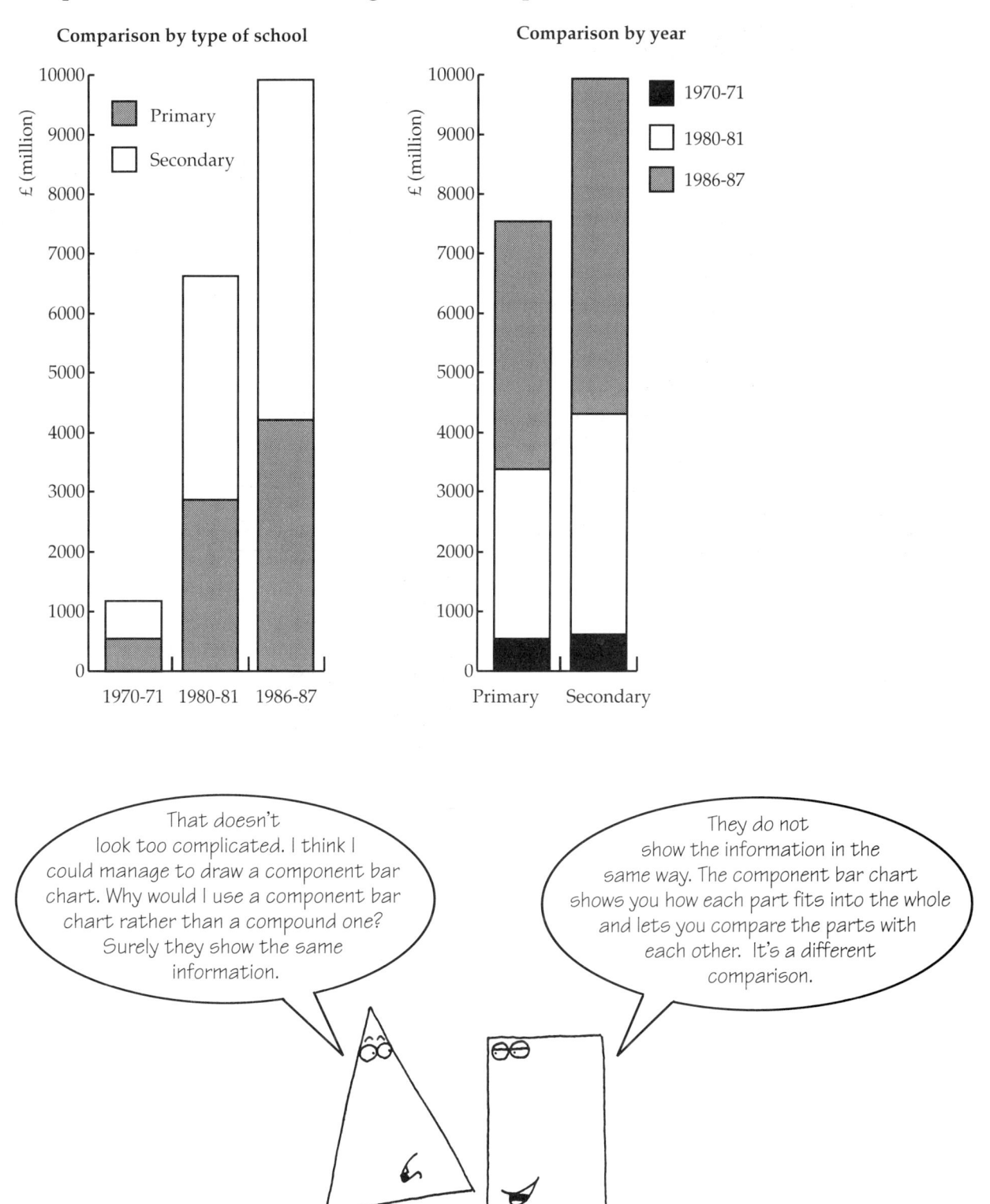

Comparison by type of school

Comparison by year

> That doesn't look too complicated. I think I could manage to draw a component bar chart. Why would I use a component bar chart rather than a compound one? Surely they show the same information.

> They do not show the information in the same way. The component bar chart shows you how each part fits into the whole and lets you compare the parts with each other. It's a different comparison.

143

▽ Try some for yourself

11. Using the data on first suppliers of drugs, draw component bar charts:

(a) by type of drug

(b) by supplier.

	Supplier		
Drug	**Friend**	**Family**	**Dealer**
Heroin	64%	8%	28%
Cannabis	78%	3%	19%
Speed	84%	6%	10%

12. Study this component bar chart illustrating the participation of 18–24 year olds in further and higher education, 1986–7.

Interpret the information displayed in the graph.

Participation of 18-24 year olds in further and higher education, 1986/87

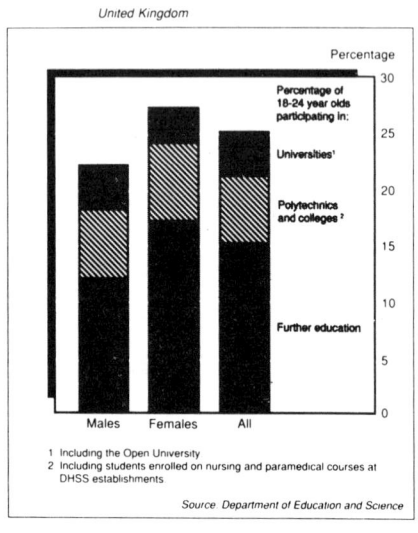

United Kingdom

[Bar chart showing Males, Females, All with components: Universities[1], Polytechnics and colleges[2], Further education]

1 Including the Open University
2 Including students enrolled on nursing and paramedical courses at DHSS establishments

Source: Department of Education and Science

Source: Social Trends 1989

144

What have you learnt about illustrating information?

There are many ways of illustrating information. I have a better idea of which method to use with particular types of data.

- [✓] I can draw a pie chart by dividing a circle into a number of parts.
- [✓] I can interpret the information given in a pie chart.
- [✓] I can construct a bar chart from a frequency table.
- [✓] I can read information from a bar chart and interpret that information.
- [✓] I can construct component and compound bar charts.
- [✓] I can identify significant features from a graph.
- [✓] I know when to use a pictogram and how to decide on the symbol to use.

Further Questions

1. Details are given below of the facilities for visitors at the Royal Botanical Gardens, Kew.

Gents' toilets	6
Ladies' toilets	5
Disabled toilets	5
Drinking fountains	6
Cafés/restaurants	3

   Draw pictograms to illustrate this information.

2. Figures for the readership of Sunday newspapers in the years 1971, 1985 and 1987 are given below. (Figures are in millions).

Newspaper	1971	1985	1987
News of the World	15.8	13.0	12.8
Sunday Mirror	13.5	10.0	9.1
People	14.4	8.9	8.1
Sunday Express	10.4	6.7	6.1
Sunday Times	3.7	4.1	3.6
The Observer	2.4	2.4	2.3
Sunday Telegraph	2.1	2.4	2.2

Source: Social Trends 1989

Illustrate the data by using compound bar charts.

3. Study the line graph and interpret the information contained in the graph.

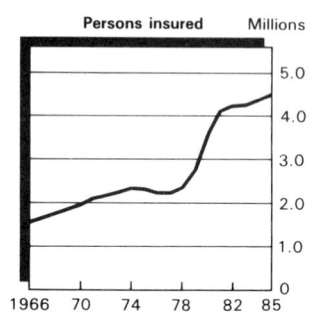

Source: Social Trends 1987

4. Numbers of students enrolled in FE institutions in the UK in 1986–7 are shown below. The numbers are in thousands.

Age	Males	Females
18 or under	375	337
19 or 20	96	75
21 or over	368	595

Source: Social Trends 1989

Use two different methods to illustrate the data.

5. **THE INCREASE IN UNIVERSITY ENROLMENT**
—PER CENT INCREASES IN STUDENT NUMBERS 1955–1967

Study this pie chart illustrating the increase in University Enrolment 1955–67.

Interpret the information displayed in the graph.

JAPAN 447
312 FRANCE
241 GERMANY
UNITED KINGDOM 230

6. **Car ownership: number of cars per 1000 of the population**

Country	1971	1981	1986
UK	224	281	323
Japan	100	211	237
USA	448	538	562

Source: Social Trends 1989

Use appropriate charts/diagrams to illustrate this data. Comment on the information displayed.

7. Study the graph showing radio and television audiences.

Interpret the information displayed in the graph.

Radio and television audiences[1] throughout the day, 1988[2]

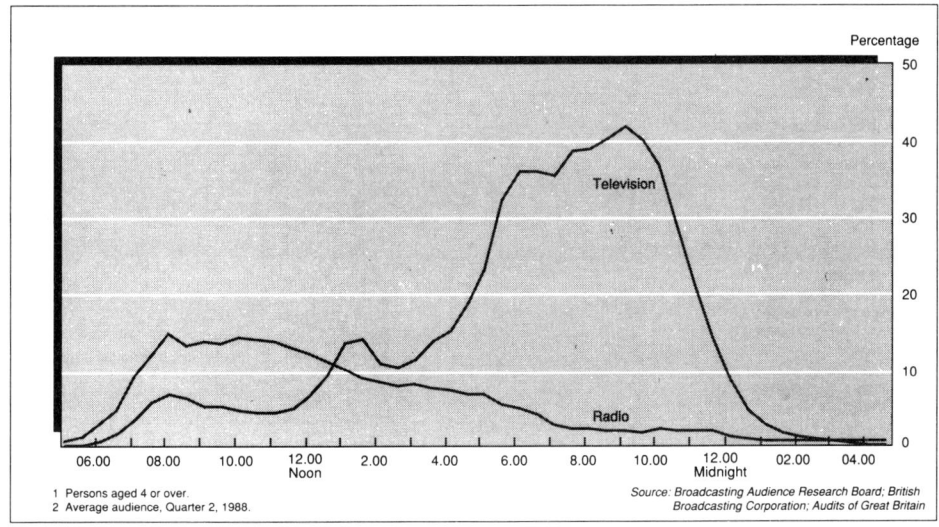

United Kingdom

1 Persons aged 4 or over.
2 Average audience, Quarter 2, 1988.

Source: Broadcasting Audience Research Board; British Broadcasting Corporation; Audits of Great Britain

8. **Readership of women's magazines** (Figures in millions)

Magazine	1971	1985
Women's Own	7.2	4.8
Woman's Weekly	4.7	3.4
Woman	8.0	3.6
Family Circle	4.4	2.7
Good Housekeeping	2.7	2.6
Woman's Realm	4.6	2.1

Source: Social Trends 1989

Use compound bar charts to illustrate this data.
Comment on your results.

147

9. Study the compound bar charts illustrating the households with VCR's.

   Interpret the information contained in the graph.

   Households with a video cassette recorder: by household type, 1983 and 1986
   Great Britain

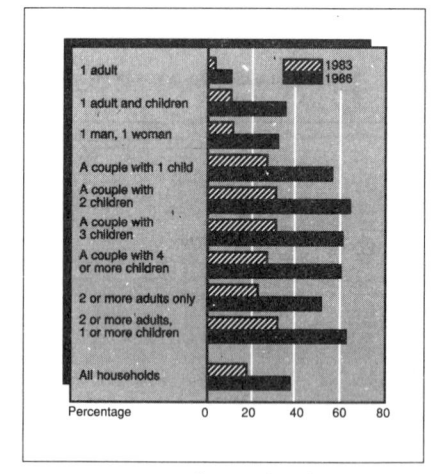

   Categories shown (top to bottom): 1 adult; 1 adult and children; 1 man, 1 woman; A couple with 1 child; A couple with 2 children; A couple with 3 children; A couple with 4 or more children; 2 or more adults only; 2 or more adults, 1 or more children; All households.
   Percentage axis: 0, 20, 40, 60, 80.

   Source: Social Trends 1989

10. Which graph would be the best way to display this data?

    Homeless households in England and Wales
    (All figures in thousands)

Type of accommodation	1984	1985	1986	1987
Bed and Breakfast	3	5	9	10
Hostels (including women's refuges)	4	5	5	5
Other accommodation	5	6	8	10

    Source: Social Trends 1989

11.

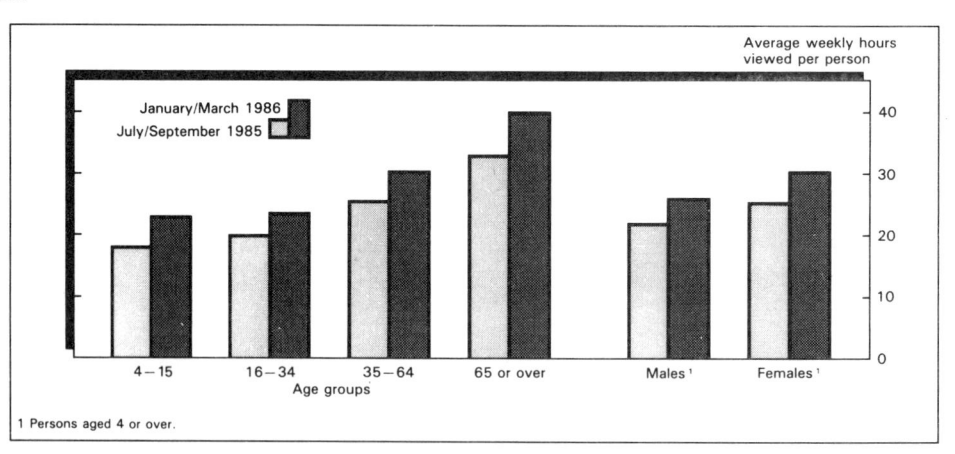

    Average weekly hours viewed per person

    January/March 1986
    July/September 1985

    Age groups: 4–15, 16–34, 35–64, 65 or over, Males[1], Females[1]

    1 Persons aged 4 or over.

    Source: Social Trends 1987

    Study the compound bar chart illustrating television viewing figures by age and sex.

    Interpret the information given in the graph.

12. Vaccination and immunisation of children in Great Britain
    Percentage of children vaccinated

Type of vaccination	1971	1976	1981	1986	1987
Poliomyelitis	80	73	82	85	87
Measles	46	45	54	71	76
Whooping Cough	78	39	46	46	73

Source: Social Trends 1989

Note In the 1970s there was widespread anxiety about the side effects of the whooping cough vaccine.

Use an appropriate graph to illustrate this data. Comment on the information contained in the graph.

13.

HIGHER EDUCATION ENROLMENT 1968-1969
AS PERCENTAGE OF AGE GROUP 20-24

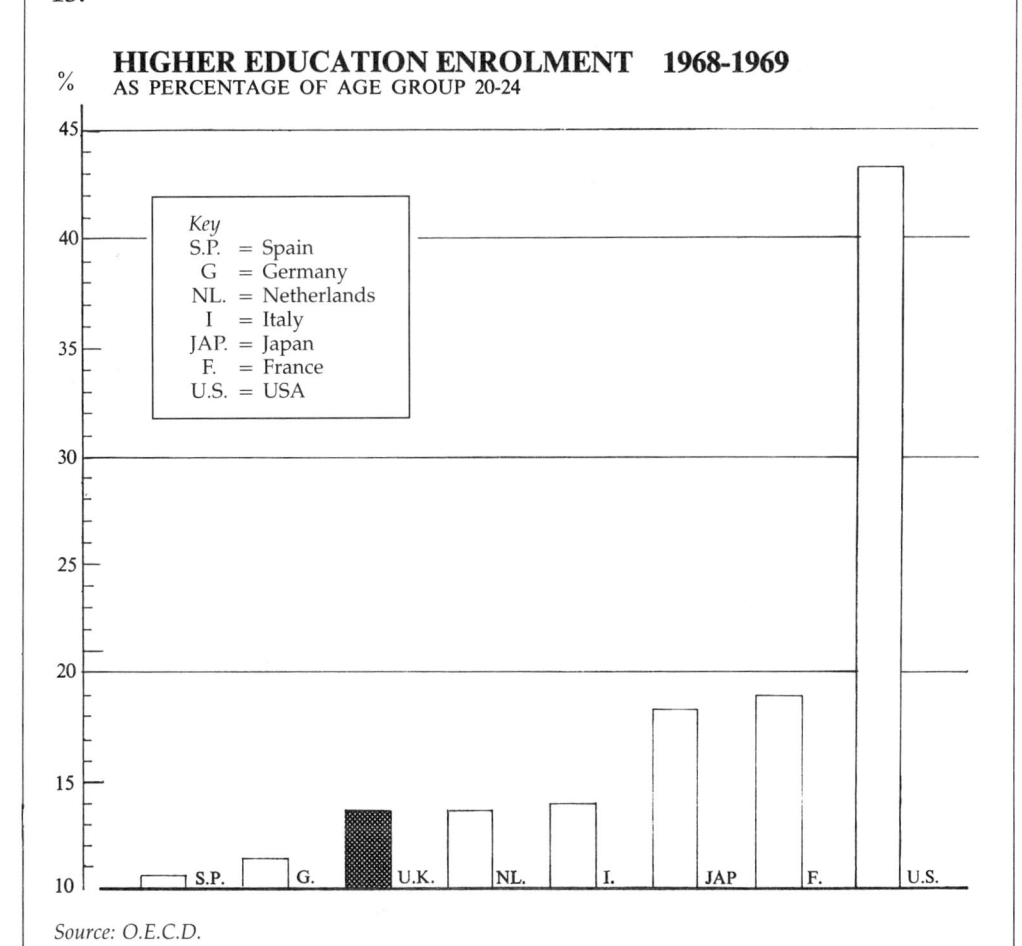

Key
S.P. = Spain
G = Germany
NL. = Netherlands
I = Italy
JAP. = Japan
F. = France
U.S. = USA

Source: O.E.C.D.

This bar chart illustrates data on the percentages of the 20–4 age group enrolling in Higher Education in 1968–9.

Comment on the information contained in the graph. Comment on the UK's position in relation to other countries.

14.

Day the accident happened

Day and date	Victims
Sunday 5 December	7
Monday 6 December	11
Tuesday 7 December	10
Wednesday 8 December	12
Thursday 9 December	16
Friday 10 December	9
Saturday 11 December	14
Total	79

Time that the accidents happened

Time	Victims
0.01am – 6am	9
6.01am – 12 noon	14
12.01am – 6pm	16
6pm – midnight	26
Not known	14
Total	79

What the victims were doing

Role	Number
Drivers	28
Passengers	13
Pedestrians	27
Cyclists	5
Motorcyclists	5
Pillion passenger	1
Total	79

The ages of the dead

Years	Number
0–10	1
11–20	16
21–30	14
31–40	8
41–50	10
51–60	6
61–70	9
71–79	11
80 plus	4
Total	79

Source: The Independent

Choose appropriate graphs or diagrams to illustrate the data contained in these four tables.

Solutions

1.

Number of hotels in each category

Category C	C C C C
Category G	G G
Category H	H H H H H H H H H H H H H H H H
Category P	P P P P P P

The category H hotels are the most common with 16. There are only two category G hotels so they must be rather exclusive.

2.

Daily hours of sunshine

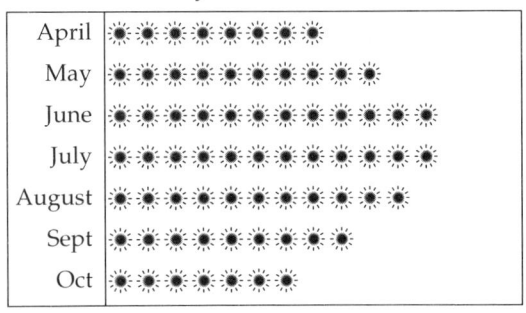

April	☀☀☀☀☀☀☀
May	☀☀☀☀☀☀☀☀☀
June	☀☀☀☀☀☀☀☀☀☀☀
July	☀☀☀☀☀☀☀☀☀☀☀
August	☀☀☀☀☀☀☀☀☀☀
Sept	☀☀☀☀☀☀☀☀
Oct	☀☀☀☀☀☀

☀ = 1 hour

3.

Students' shoe sizes

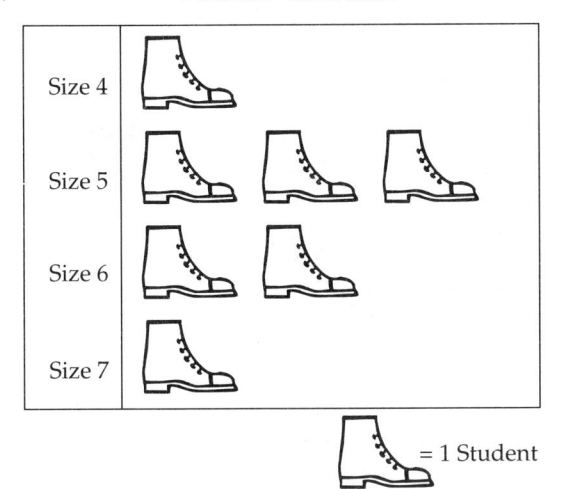

Size 4	🥾
Size 5	🥾 🥾 🥾
Size 6	🥾 🥾
Size 7	🥾

🥾 = 1 Student

4.

GNVQ tutor group

| Male | 👨 👨 👨 👨 👨 👨 👨 👨 |
| Female | 👩 👩 👩 👩 👩 👩 |

👨 = 1 male student 👩 = 1 male student

5.

Category	Total	Angle	Rounded angle
C	4	$\frac{4}{28} \times 360° = 51.42°$	52°
G	2	$\frac{2}{28} \times 360° = 25.71°$	26°
H	16	$\frac{16}{28} \times 360° = 205.71°$	206°
P	6	$\frac{6}{28} \times 360° = 77.14°$	77°
	28		360°

(Pie chart with sectors labelled C, G, P, H)

6. (a)

Weekly Sales — £ (thousands) vs dates from 7 December to 23 February. Values rise from 17 on 7 Dec to a peak of 31 on 5 Jan, then drop sharply to around 11 on 19 Jan and decline to 5 by 23 Feb.

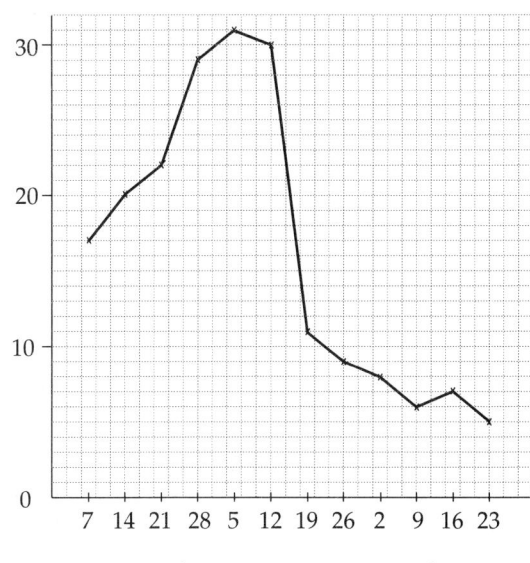

(b) This graph shows a rise in sales in the period leading up to Christmas. Sales peaked at the beginning of January, probably due to the January Sales. The figures dropped dramatically from January 19th onwards. This is probably when the sales ended.

7. (a)

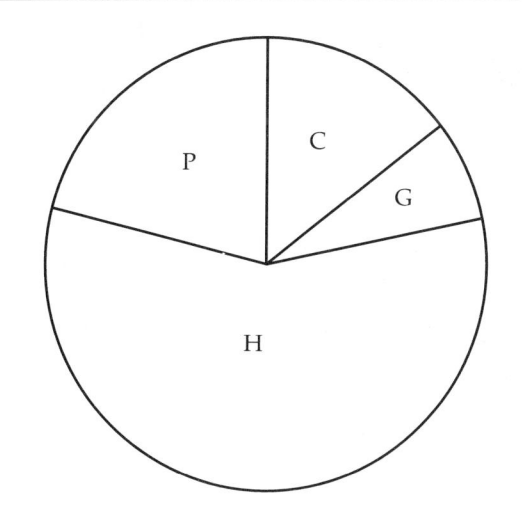

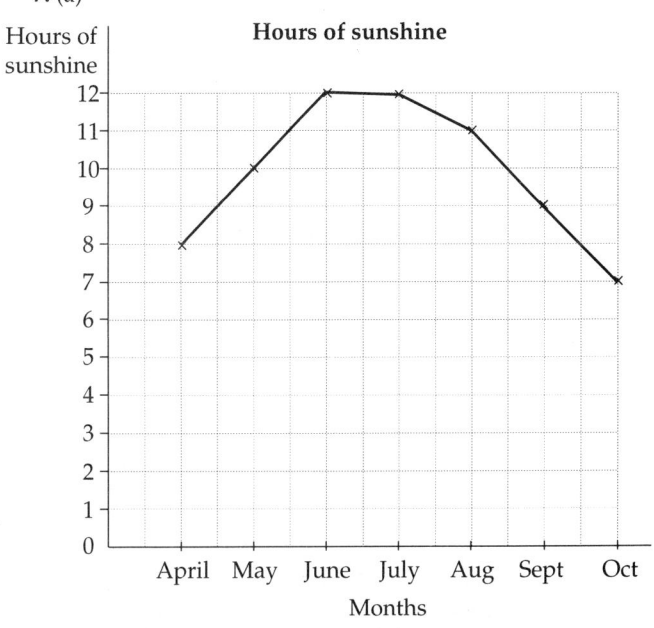

Hours of sunshine by month: April 8, May 10, June 12, July 12, Aug 11, Sept 9, Oct 7.

(b) The hottest months are June, July and August so the sunseekers should take their holidays then. For those who prefer less exposure to the sun, the months of April, May, September and October will be more suitable.

8.

Bar chart showing hotels in each category

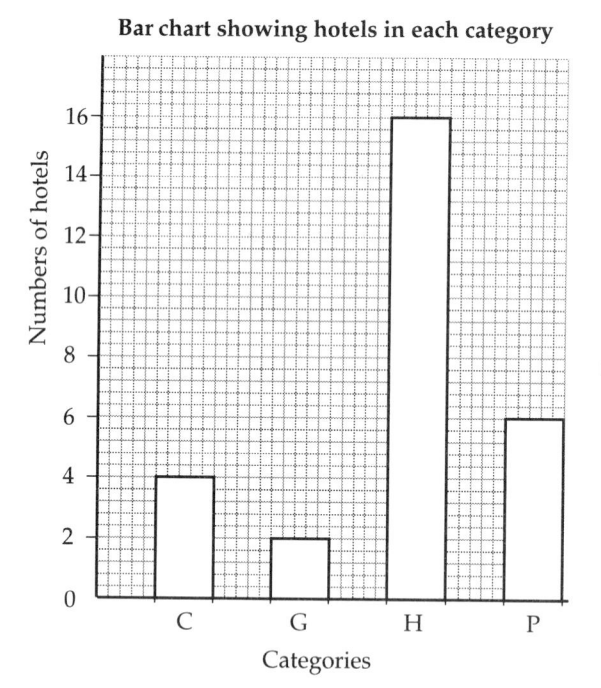

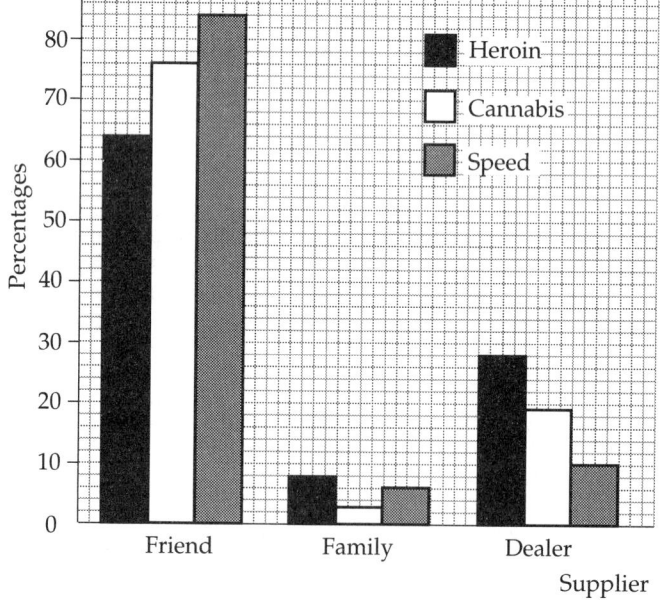

9.

Bar chart showing drug usage

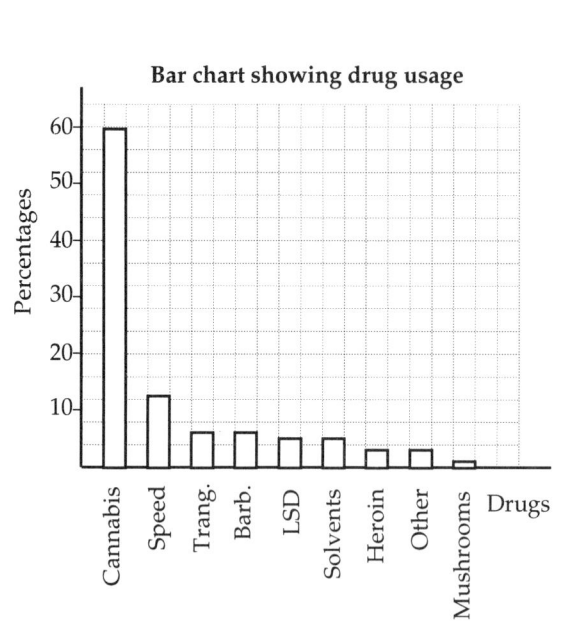

The striking factor illustrated by this chart is the use of cannabis. It is the most frequently used drug. Speed has only 12% usage. This may be because cannabis is thought to be non-addictive, all the other drugs are known to be addictive.

10.

Drug suppliers, by drug

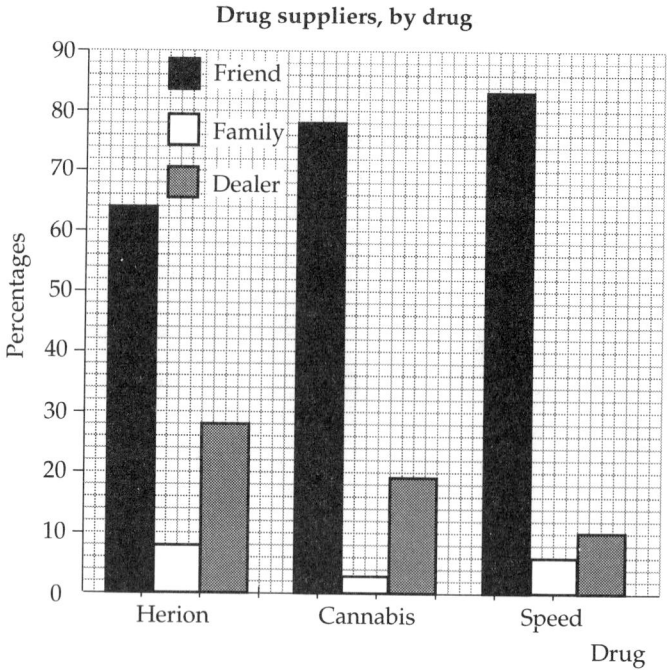

Drug suppliers, by supplier

152

11. (a)

Drug supply, by drug

Legend: Dealer, Family, Friend

(b)

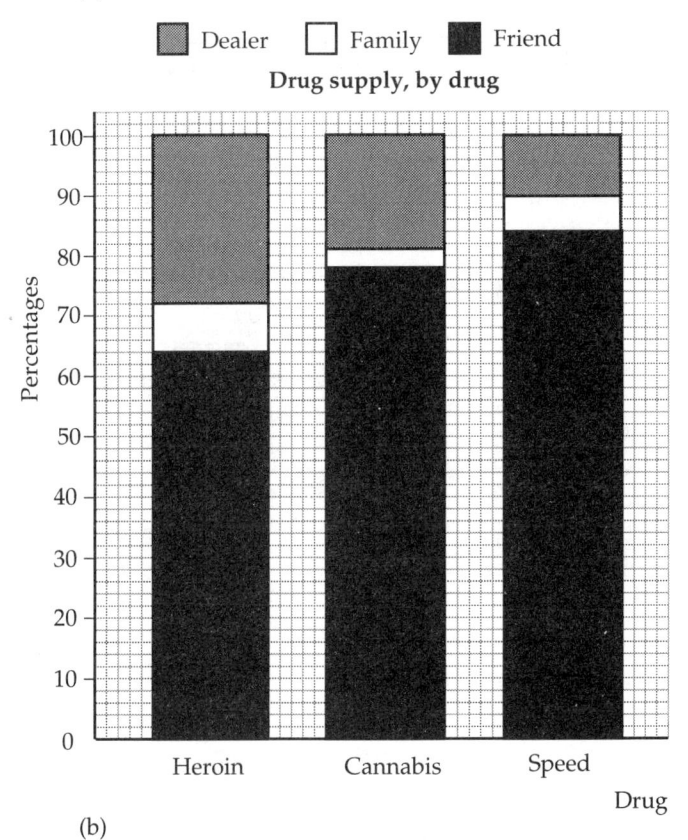

Drug supply, by supplier

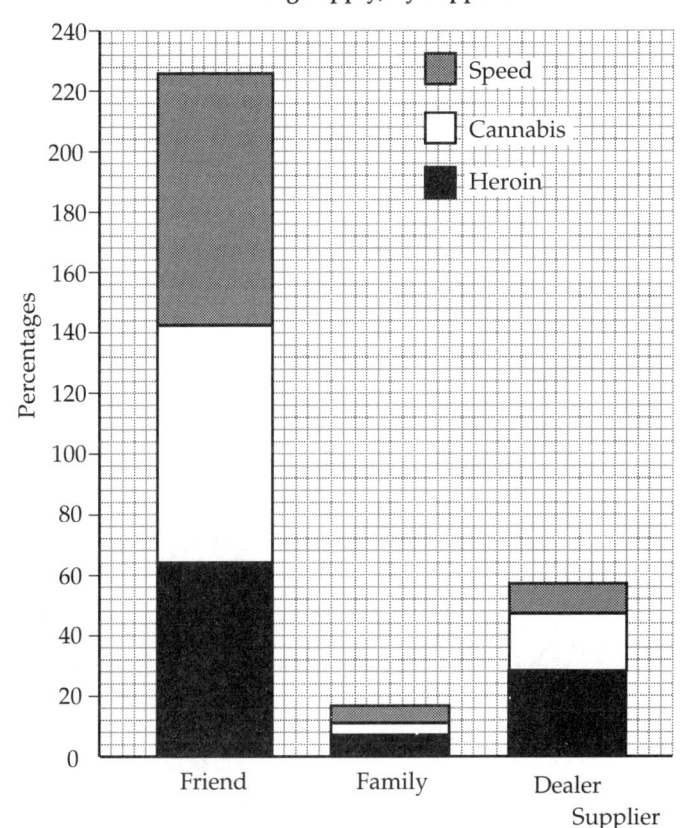

Legend: Speed, Cannabis, Heroin

12. The chart shows that approximately 28% of females aged 18–24 were in further or higher education compared to only 23% of males. More female students attended further education colleges (approx. 17%). The numbers attending polytechnics were almost equal with slightly more females than males. The figures for universities show a slight majority of male students.

Female students are more likely to go into further education but approximately the same number of both sexes enter higher education.

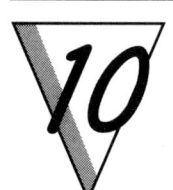

10 Measuring Information

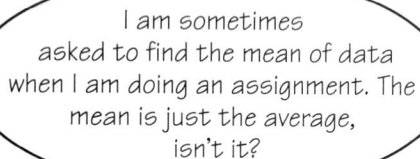

> I am sometimes asked to find the mean of data when I am doing an assignment. The mean is just the average, isn't it?

> An average is just a single figure that represents the whole group of data. We use several types of average without really understanding what each means.

> I thought an average was just the sum of the numbers divided by the total. Now you tell me that an average can be something different. Could you tell me more about averages?

> There are several types of average in use, but you only need to know about the three we use most often. I'm sure you have heard of these before: the mean, the median and the mode.

> I think I have heard of all of those but I am not very confident about finding them or knowing when to use them. Some revision would be very useful.

The Mean

This is also called the **arithmetic mean**. The mean is what you think of when you are asked to find the average. It's the most commonly used of the measures of average. It is easy to calculate and easy to understand.

Calculating the mean

1. A batsman's scores in 6 cricket matches have been recorded and his mean score is to be calculated.

Scores: 26,   52,   12,   37,   75,   0

How do we calculate the mean?

We add all the scores together to find his total score for all 6 matches, then we divide that total by 6, the total number of matches.

$$26 + 52 + 12 + 37 + 75 + 0 = 202$$

$$\frac{202}{6} = 33.67$$

So the mean score is 33.67 runs.

2. A survey was carried out to find the number of children per family in a sample of 10 families.

The results were:

3,   2,   2,   5,   1,   1,   2,   3,   2,   3

Total number of children $= 3 + 2 + 2 + 5 + 1 + 1 + 2 + 3 + 2 + 3 = 24$

Total number of families $= 10$

Mean $= \dfrac{24}{10} = 2.4$

The mean number of children per family is 2.4.

$1 + 1 + 0.4 = 2.4$

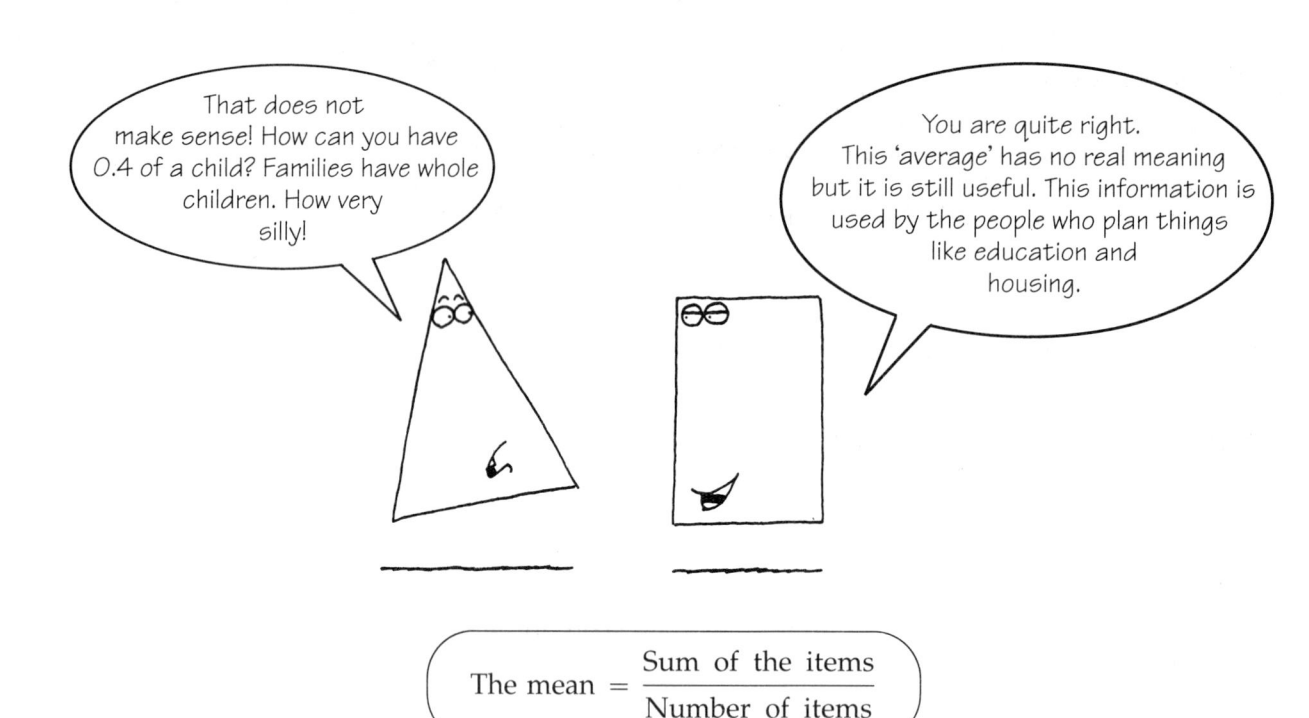

That does not make sense! How can you have 0.4 of a child? Families have whole children. How very silly!

You are quite right. This 'average' has no real meaning but it is still useful. This information is used by the people who plan things like education and housing.

$$\text{The mean} = \frac{\text{Sum of the items}}{\text{Number of items}}$$

Try some for yourself

1. A GNVQ student gained the following marks in a series of maths tests.

      45   30   60   50   65   75

What was the student's mean mark?

2. The average daily hours of sunshine in Tunisia during the summer season are given.

April	8	August	11
May	10	September	9
June	12	October	7
July	12		

Find the mean monthly hours of sunshine during the summer season.

Grouped data

That's very easy. All I need to do is to add up all the numbers and divide by the total number of numbers. Finding the mean is a piece of cake!

Don't get too carried away. You have only worked out the mean of a list of numbers so far. What will you do with data presented like this?

Number of children	Frequency
1	2
2	4
3	3
4	0
5	1
Total	10

> You have added up all the frequencies so all I need to do is to add up the number of children. The total is 15 so I divide that by the total frequency to find the mean!

$$15 \div 10 = 1.5$$

The mean number of children is 1.5

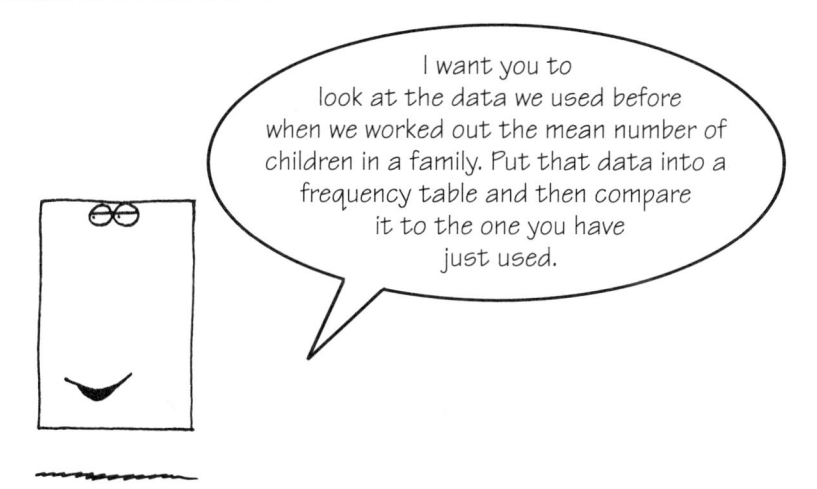

> I want you to look at the data we used before when we worked out the mean number of children in a family. Put that data into a frequency table and then compare it to the one you have just used.

Number of children	Tally	Frequency
1	11	2
2	1111	4
3	111	3
4		0
5	1	1
	Total	10

157

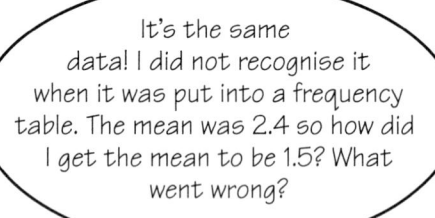

It's the same data! I did not recognise it when it was put into a frequency table. The mean was 2.4 so how did I get the mean to be 1.5? What went wrong?

You did not stop to think exactly what the frequency table shows you. The frequency table shows you that two families have one child each — that's two children in total.

That means that I must multiply the number of children by the frequency figure each time. If I add those totals up I bet I shall get a grand total of 24. I'll try it.

Number of children	Frequency	Total
1	2	2
2	4	8
3	3	9
4	0	0
5	1	5
Total	10	24

> That's better! I now have a total of 24 children in 10 families. When I divide 24 by 10 I shall have an answer of 2.4, which is the same as we had earlier, so I know I am right.

> You have worked out this method for yourself which is very good. I think you will remember the method now. Perhaps some examples will help.

Try some for yourself

3. Find the mean number of four-poster beds in these hotels.

Number of four-posters	Frequency
1	3
2	7
3	1
4	0
5	1

4. Find the mean price per person in this sample of hotel prices.

Price (£)	43	49	54	58	59	64	74	79
Frequency	5	2	6	3	1	2	2	1

> All these calculations involve a lot of arithmetic. I use my calculator but I have to write down all the numbers first before I can enter them and find the total. Surely, there must be an easier way of doing the calculations?

> You need to use your calculator's memory to store the running total as you go along. Then, when you have entered all the numbers you can get the final total from the memory.

> That sounds wonderful but how do I do it? I have used the memory for fractions and bits in brackets but this sounds more complicated. I think I need help.

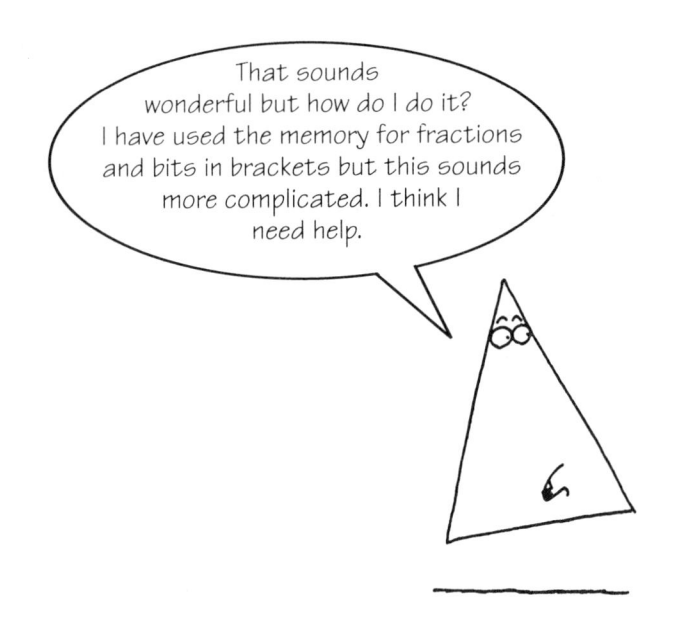

The keys you will use are:

[0][x→m] or [0][Min]	This enters 0 into the memory, clearing any earlier numbers away.
[Min] or [x→m]	Stores the current number in the memory.
[M+]	Adds the current number to the memory.
[MR] or [RM]	Recalls the total from the memory.

As the current number is added to the memory, the total is automatically increased. When the last number has been added to the memory, the final total will be stored in the memory.

A worked example is the best way to see how the calculator's memory works.

Number	1	2	3	4	5
Frequency	2	4	3	0	1

	Display
[0][x→m] or [0][Min]	0
[1][×][2][=][M+]	2
[2][×][4][=][M+]	8
[3][×][3][=][M+]	9
[4][×][0][=][M+]	0
[5][×][1][=][M+]	5
[MR] or [RM]	24

> That is wonderful! It's a much easier way to find the total of a frequency table. I shall have to practice using my calculator's memory.

Try some for yourself

5. The frequency table shows the length of time taken by 35 students on a GNVQ course, to travel to college each day. Find the mean travelling time.

Time (mins)	5	10	15	20	25	30	40	45	50
Frequency	4	1	0	10	9	2	1	7	1

6. The frequency table shows the number of absences recorded for the GNVQ students during their first term.

Find the mean number of absences.

Absences	0	1	2	3	4	5	6
Frequency	18	6	5	2	0	3	1

> I feel very confident about finding the mean of data. Now I can use my calculator properly the calculations are easy. What about the other common averages you said I need to know about? Tell me about the median and the mode.

> The median just identifies the middle item in a list — it's very simple. The mode is the most popular item — the one which occurs most often. Both averages have their special uses. I'll give you a brief explanation of them.

The median

The median is the average that refers to the number in the middle of a list of numbers.

$$2 \quad 1 \quad 3 \quad 4 \quad 5 \quad 7 \quad 1$$

To find the median we must first arrange the numbers in numerical order.

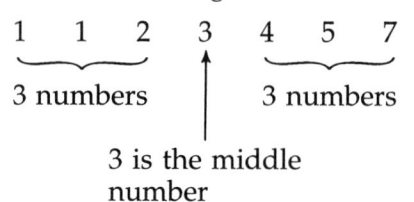

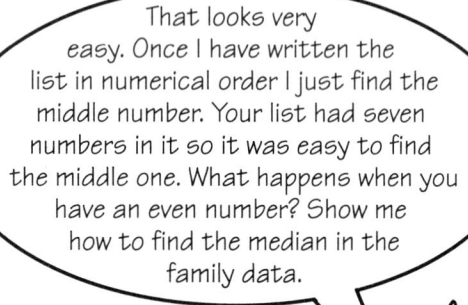

$$\underbrace{1 \quad 1 \quad 2}_{3 \text{ numbers}} \quad 3 \quad \underbrace{4 \quad 5 \quad 7}_{3 \text{ numbers}}$$

3 is the middle number

> That looks very easy. Once I have written the list in numerical order I just find the middle number. Your list had seven numbers in it so it was easy to find the middle one. What happens when you have an even number? Show me how to find the median in the family data.

> Finding the median when the list contains an even number of values is more complicated. I'm pleased that you spotted the difficulty so soon. The solution is quite straightforward.

Family data

$$3 \quad 2 \quad 2 \quad 5 \quad 1 \quad 1 \quad 2 \quad 3 \quad 2 \quad 3$$

Rearranging in ascending numerical order:

$$1 \quad 1 \quad 2 \quad 2 \quad 2 \underset{\text{middle}}{\uparrow} 2 \quad 3 \quad 3 \quad 3 \quad 5$$

The 'middle' lies between the fifth and sixth numbers. So the median is the mean of these two numbers.

$$\frac{2+2}{2} = 2$$

The median is 2.

▽ Try some for yourself

7. A GNVQ student gained the following marks in a series of maths tests.

$$45 \quad 30 \quad 60 \quad 50 \quad 65 \quad 70$$

What was the median mark?

8. The average daily hours of sunshine in Tunisia during the summer season are given in this table.

April	May	June	July	Aug	Sept	Oct
8	10	12	12	11	9	7

What was the median number of hours?

The mode

The mode is the average that identifies the most popular item or number in a list.

$$2 \quad 1 \quad 3 \quad 4 \quad 5 \quad 7 \quad 1$$

We see that 1 is the number occurring most often so the **mode** is 1.

Family data

Number	1	2	3	4	5
Frequency	2	4	3	0	1

Here 2 is the number that occurs most often. The mode is 2.

This tells us that the most common size of family has 2 children. The mean size was 2.4 children.

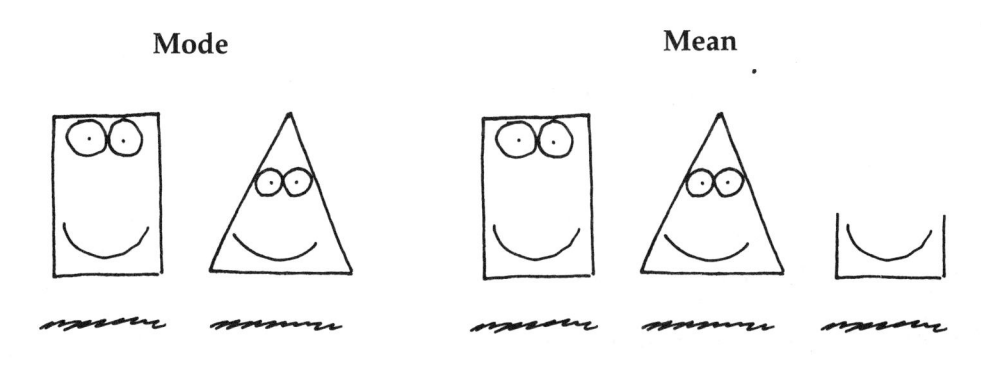

     Mode   Mean

  2 children   2.4 children

This demonstrates the different answers we get when we work out 'averages'. We must be precise about the average that is used.

163

▽ Try some for yourself

9. Sales of stationery in a college shop are given in the table. Find the mode.

Weekly sales

A4 pads	40
Black pens	35
Clear pockets	30
Project files	18
Green pens	2
Coloured paper	0

10. Find the mode in this list of shoe sizes.

4, 5, 5, 6, $6\frac{1}{2}$, 6, 7, 6, 3, 6

The range

*I'm glad we have sorted out those averages. I have learned about them all before but I used to get them confused. I think I'm much clearer about them now. Now, just one more question. What is the **range**?*

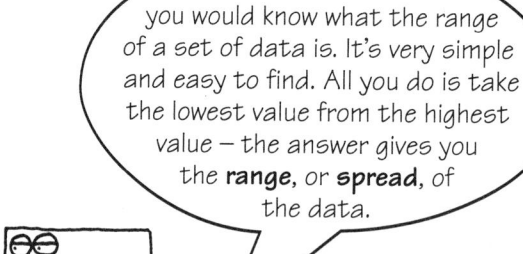

*I wondered if you would know what the range of a set of data is. It's very simple and easy to find. All you do is take the lowest value from the highest value – the answer gives you the **range**, or **spread**, of the data.*

The GNVQ students maths test scores were:

30, 45, 60, 50, 65, 70

The highest score was 70.
The lowest score was 30.

$$70 - 30 = 40$$

The range of the marks was 40.

The hours of sunshine in Tunisia were:

8, 10, 12, 12, 11, 9, 7

$$12 - 7 = 5$$

The range of hours of sunshine was 5.

That looks so simple and it's very easy to work out. I'm sure I shall have no trouble with the range. I suppose it must be useful or I would not need to know how to find it. What use is it?

The range is one of the 'measures of dispersion', in other words it tells you how spread out the data is. You usually use it with the mean to see how close the data is to the mean. Let me give you some examples of the use of the range.

Car manufacturers promote a particular model which has a number of variations which are sold in a *range* of prices.

When you order goods you are given a *range* of time for the delivery e.g. 8–10 days.

Manufacturers will sample the output of a production line at regular intervals. The sample mean and *range* are used to allow the quality controller to check the variation of output.

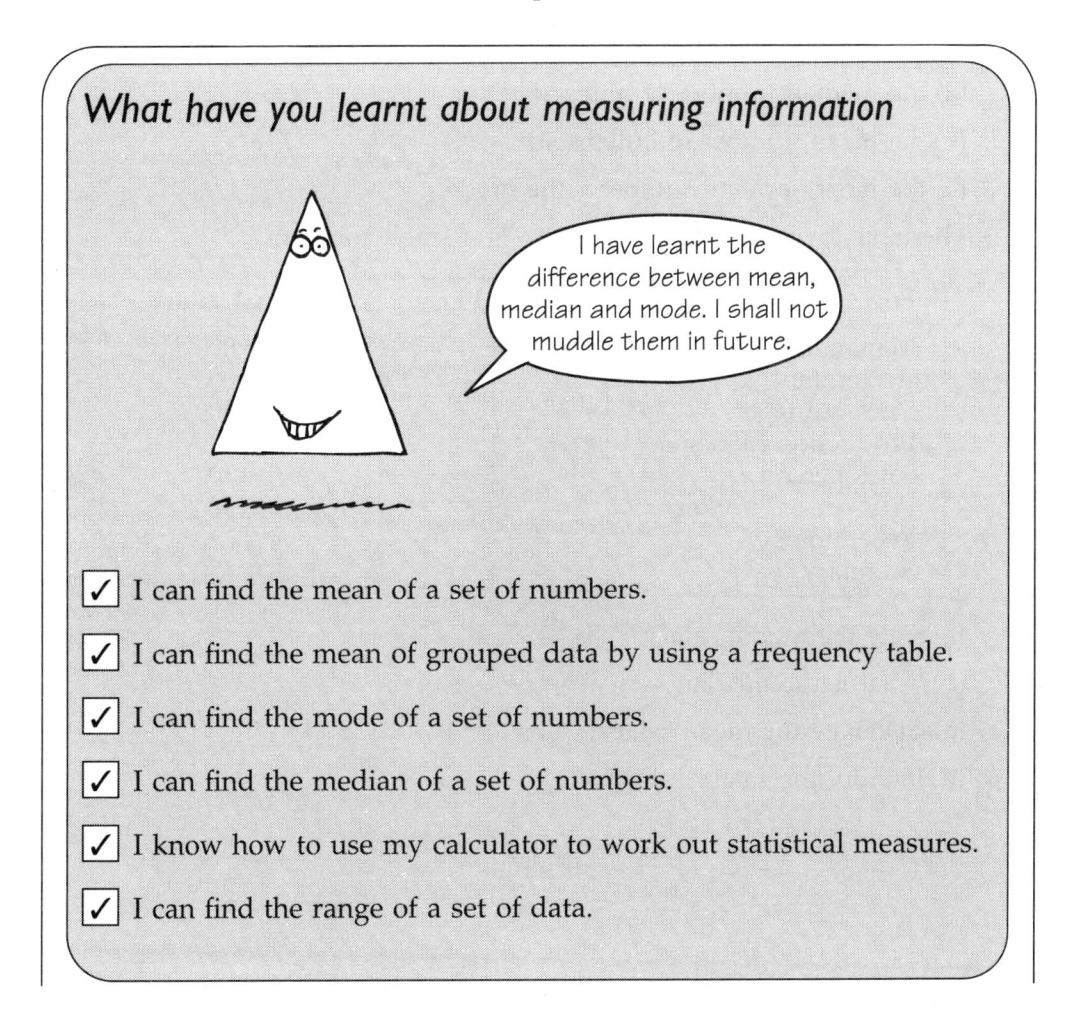

What have you learnt about measuring information

I have learnt the difference between mean, median and mode. I shall not muddle them in future.

- [✓] I can find the mean of a set of numbers.
- [✓] I can find the mean of grouped data by using a frequency table.
- [✓] I can find the mode of a set of numbers.
- [✓] I can find the median of a set of numbers.
- [✓] I know how to use my calculator to work out statistical measures.
- [✓] I can find the range of a set of data.

Further Questions

1. You are given information about prices in a number of hotels. Using this data you are required to:

    (a) Calculate the mean price

    (b) Calculate the median price

    (c) Identify the modal price

    (d) State the price range.

    £49, £49, £49, £45, £49, £54, £79, £54, £43, £45, £49, £54, £58, £49, £64, £64, £45, £45, £43, £58, £54, £58, £74, £45, £74, £59, £54, £54

2. One day a garden centre sold the following number of bags of snowdrop bulbs.

Number of bulbs per bag	10	15	25	50	100
Number of bags sold	1	3	5	12	9

    Use this information to find:

    (a) the median number of bulbs sold

    (b) the mean number of bulbs sold

    (c) the most popular number – the mode.

3. The table gives the salary structure of a small company.

Post	Number	Annual salary
Managing Director	1	£45 000
Accountant	1	£30 000
Sales Manager	1	£19 500
Production Manager	1	£19 000
Sales person	2	£ 9 500
Production Worker	9	£ 8 500
Receptionist	1	£ 4 250
Secretary	4	£ 6 550

    (a) State the modal salary?

    (b) What is the median salary?

    (c) Calculate the mean salary?

    (d) Identify the salary range.

4. The following figures give the annual totals (in thousands) of new housing in the years from 1981–7.

Year	1981	1982	1983	1984	1985	1986	1987
Totals	207	182	207	218	203	208	212

Source: Social Trends 1989

Find the mean, median and mode.

Solutions

1. $45 + 30 + 60 + 50 + 65 + 75 = 325$

$$\frac{325}{6} = 54.17 \text{ to 2 d.p.}$$

Mean mark = 54.17 to 2 d.p.

2. $8 + 10 + 12 + 12 + 11 + 9 + 7 = 69$

$$\frac{69}{7} = 9.86 \text{ to 2 d.p.}$$

Mean monthly hours of sunshine = 9.86 h to 2 d.p.

3.
$1 \times 3 = 3$
$2 \times 7 = 14$
$3 \times 1 = 3$
$4 \times 0 = 5$
$5 \times 1 = \underline{5}$
25

$$\frac{25}{5} = 5$$

Mean number of four-poster beds = 5

4.
$43 \times 5 = 215$
$49 \times 2 = 98$
$54 \times 6 = 324$
$58 \times 3 = 174$
$59 \times 1 = 59$
$64 \times 2 = 128$
$74 \times 2 = 148$
$79 \times 1 = \underline{79}$
25

$$\frac{1225}{22} = £55.68$$

Mean price per person = £55.68

5. Total time 920
Total frequency = 35

Mean = 26.29 to 2 d.p.

Mean travelling time = 26.29 min to 2 d.p.

6. Total absences = 45
Total frequency = 35

Mean absences = 1.23 to 2 d.p.

7. 30 45 50 60 65 70

Middle number

$$\text{Median} = \frac{50 + 60}{2} = \frac{110}{2} = 55$$

8. 7 8 9 10 11 12 12

middle

Median = 10

9. 0 2 18 30 35 40

largest number

Mode = 40   A4 pads sell best.

10. Mode = 6

167

11 Probability

> I see that I need to know a little bit about probability. I don't understand why. It's all about tossing coins or throwing dice – gambling really. What use is it to anyone who is not a gambler?

> It's true that for nearly 200 years the laws of probability were used by gamblers, after all, that's why they were developed. Now, probability is an incredibly useful tool for business, government and academics.

> I certainly had no idea that probability actually had practical applications. I'd like to know what sort of use all these people make of these strange laws.

> I'll give you some examples. Insurance companies use it to work out the likely number of claims they should expect each year. Government use it to predict trends in population. Scientists use it to predict results. Businesses use it to predict the likely result of an investment. We shall be looking at the simple examples.

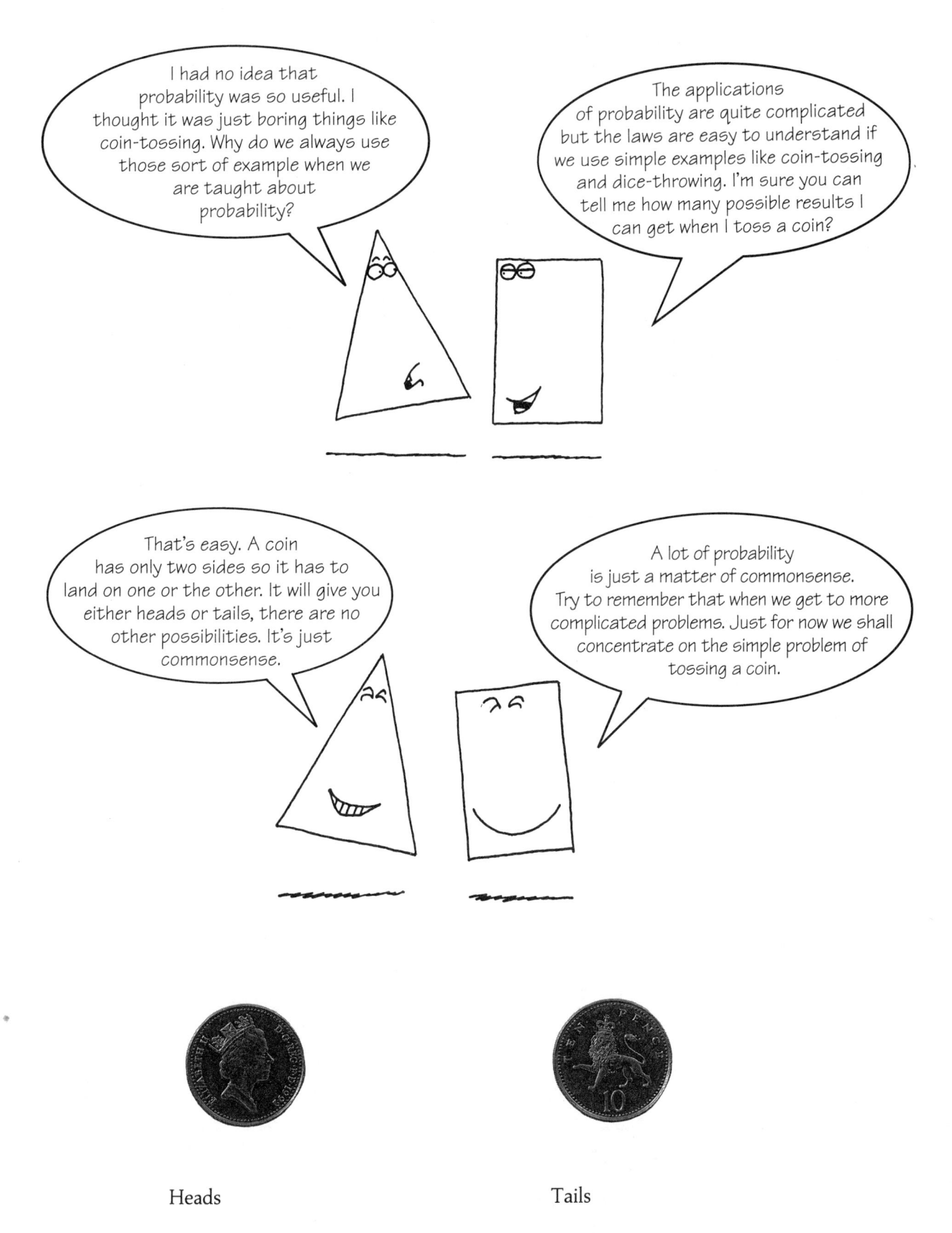

I had no idea that probability was so useful. I thought it was just boring things like coin-tossing. Why do we always use those sort of example when we are taught about probability?

The applications of probability are quite complicated but the laws are easy to understand if we use simple examples like coin-tossing and dice-throwing. I'm sure you can tell me how many possible results I can get when I toss a coin?

That's easy. A coin has only two sides so it has to land on one or the other. It will give you either heads or tails, there are no other possibilities. It's just commonsense.

A lot of probability is just a matter of commonsense. Try to remember that when we get to more complicated problems. Just for now we shall concentrate on the simple problem of tossing a coin.

Heads　　　　　　　　　　　Tails

A coin has two sides so it must give either heads or tails. There are two possible outcomes.

There is only one way the coin can land showing heads.

$$\text{Probability of heads} = \tfrac{1}{2}$$

$$\text{Probability of tails} = \tfrac{1}{2}$$

$$\text{Probability} = \frac{\text{Number of ways something can happen}}{\text{Total number of possible outcomes}}$$

Let's turn our attention to dice-throwing.

We talk of one die but two dice.

A die has six faces, each face shows a different number.

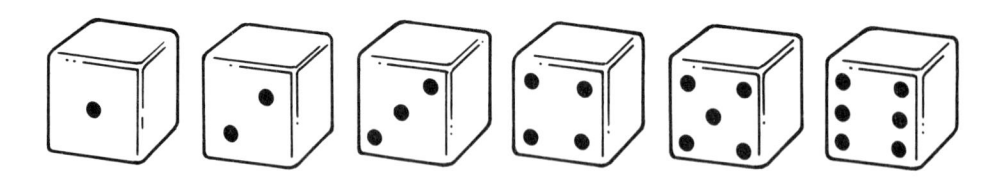

There are six possible outcomes so the total number of possible outcomes will be 6.

How many ways can we throw a 3 with a single throw. There is only one possible way to throw a 3.

$$\text{Probability of a 3} = \tfrac{1}{6}$$

$$\text{Probability} = \frac{\text{Number of possible ways to throw a 3}}{\text{Total number of possible outcomes}}$$

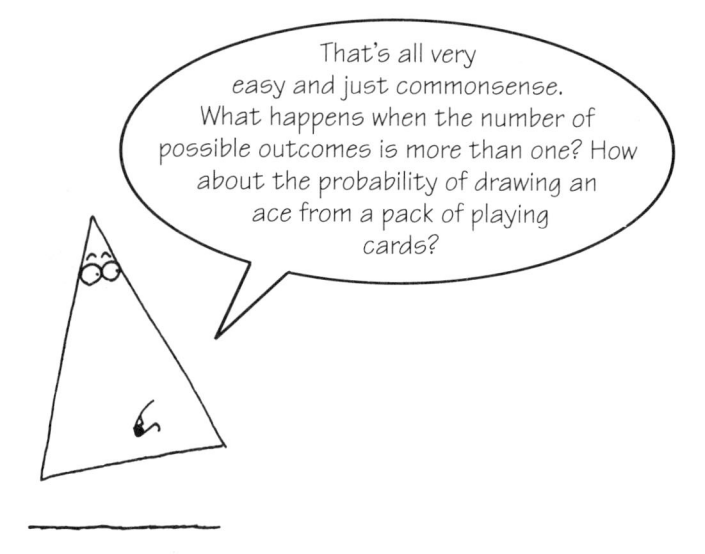

> That's all very easy and just commonsense. What happens when the number of possible outcomes is more than one? How about the probability of drawing an ace from a pack of playing cards?

A pack of playing cards contains a total of 52 cards, in four suits of 13 cards each.

Ask yourself this question:

How many aces are there in a pack of cards?

There are 4 aces.

How many cards in the pack?   52

$$\frac{\text{Number of aces}}{\text{Total number of cards}} = \frac{4}{52}$$

$$\text{Probability of an ace} = \frac{4}{52} = \frac{1}{13}$$

Let's try another card problem.

What is the probability of drawing a jack or a queen?

How many jacks are there in a pack?   4
How many queens are there in a pack?   4
How many jacks and queens?   8

$$\text{Probability of jack or a queen} = \frac{8}{52} = \frac{2}{13}$$

▽ Try some for yourself

1. What is the probability of a vowel when you pick a letter at random from:

    (a) the alphabet

    (b) the word 'probability'

    (c) the words 'Application of Number'?

2. What is the probability of drawing these cards from a pack of 52 cards?

    (a) A queen

    (b) A black three

    (c) A two or a three?

I think I can work out probabilities with only one coin, one die and one pack of cards. Lots of games use two dice or two packs of cards. How do I work out those probabilities?

When we look at more complicated combinations we need to use tables and diagrams to work out the possible outcomes. I expect you have used tables before, so we shall look at those first.

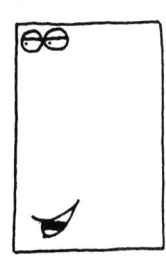

Probability tables

How many possible outcomes are there when two coins are tossed?

	Coin 1	Coin 2
Outcome 1	H	H
Outcome 2	H	T
Outcome 3	T	H
Outcome 4	T	T

There are 4 possible outcomes.

How many ways can you throw two heads?   1

$$\text{Probability of two heads} = \tfrac{1}{4}$$

$$\text{Probability of head and tail} = \tfrac{2}{4} = \tfrac{1}{2}$$

In the game of Dungeons and Dragons 3-faced dice are used. To work out the total number of possible outcomes we can use a table.

		First die		
		1	2	3
Second die	1	1 + 1	1 + 2	1 + 3
	2	2 + 1	2 + 2	2 + 3
	3	3 + 1	3 + 2	3 + 3

		First die		
		1	2	3
Second die	1	2	3	4
	2	3	4	5
	3	4	5	6

We see that there are 9 possible outcomes.

Let's find the probability of scoring a total of 4.

From the table we can see that there are three ways of scoring 4.

$$\text{Probability of 4} = \tfrac{3}{9} = \tfrac{1}{3}$$

▽ Try some for yourself

3. Produce a table showing all possible outcomes when throwing two six-sided dice.

   Use the table to find:

   (a) the probability of throwing a total of 12

   (b) the probability of throwing a total of 7

   (c) the probability of throwing any number *less* than 7.

4. Produce a table to show all possible outcomes when tossing three coins.

   What is the probability of throwing no heads?

   What is the probability of throwing one head and two tails?

> These tables are very useful and easy to produce. There would be problems if you were using 20-sided dice as you do in Dungeons and Dragons. That would mean a 20 × 20 table with 400 entries.

> When we are faced with the large tables we usually use another technique. We use special diagrams called **tree diagrams**. The tree has branches representing probabilities.

Tree diagrams

Let's go back to tossing a single coin.

I can toss a coin which will land on heads or tails so I have two branches.

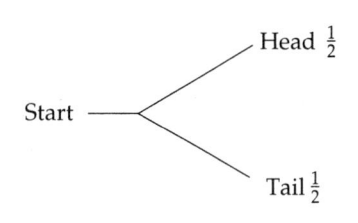

Start —< Head $\frac{1}{2}$ / Tail $\frac{1}{2}$

Let's add a second coin. Our tree has branches and twigs.

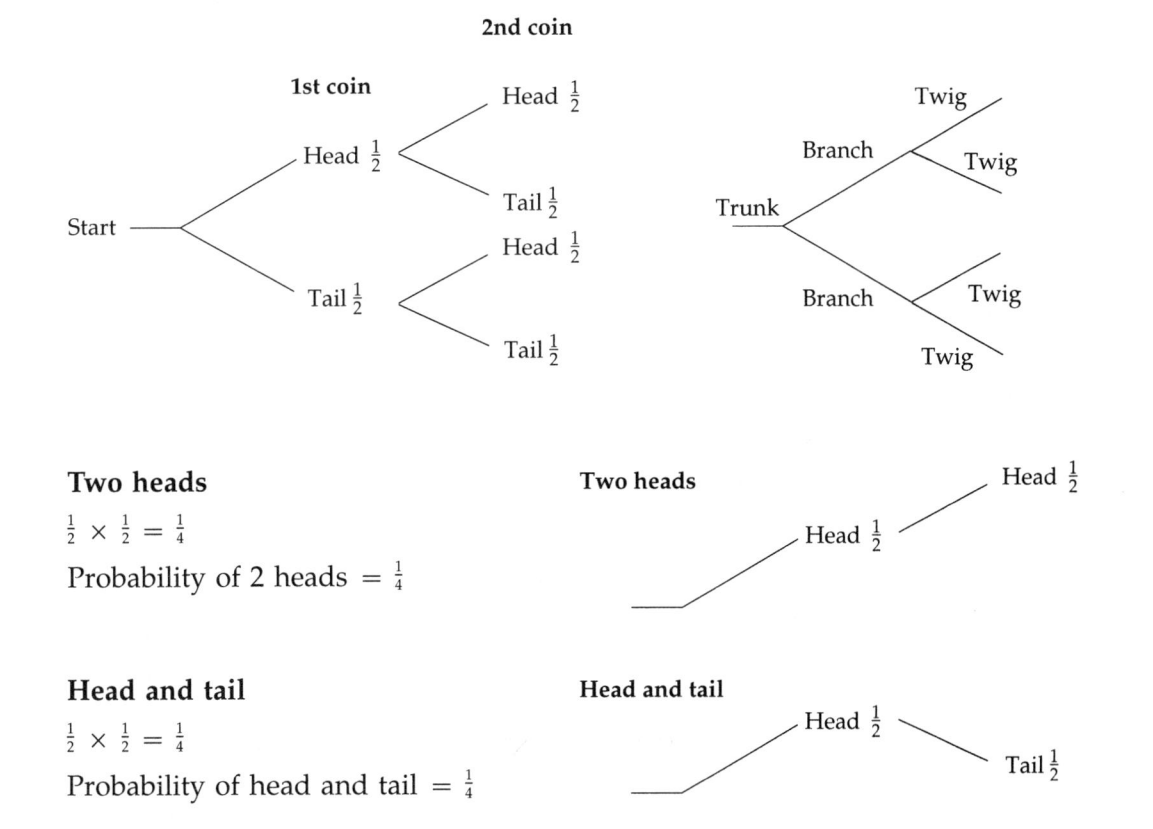

Two heads

$\frac{1}{2} \times \frac{1}{2} = \frac{1}{4}$

Probability of 2 heads = $\frac{1}{4}$

Head and tail

$\frac{1}{2} \times \frac{1}{2} = \frac{1}{4}$

Probability of head and tail = $\frac{1}{4}$

Tail and head

$\frac{1}{2} \times \frac{1}{2} = \frac{1}{4}$

Probability of tail and head = $\frac{1}{4}$

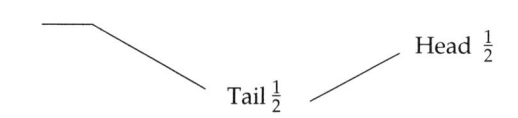

Tail and head

Tail $\frac{1}{2}$ — Head $\frac{1}{2}$

Two tails

Two tails

$\frac{1}{2} \times \frac{1}{2} = \frac{1}{4}$

Probability of 2 tails = $\frac{1}{4}$

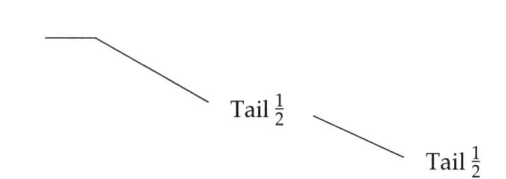

Tail $\frac{1}{2}$ — Tail $\frac{1}{2}$

> Notice that the probabilities add up to a total of **1**.

To find the probability of a head and tail we add together the probability of head followed by tail and the probability of tail followed by head.

$$\frac{1}{4} + \frac{1}{4} = \frac{2}{4} = \frac{1}{2}$$

To find the probabilities of each path we start at the trunk, follow the branch and end at the twig.

To find the probability of several paths we add their probabilities together.

> Multiply from trunk to twig.
> Add from twig to twig.

> That's easy to remember. It will also remind me to draw a tree. Now I think I need some practice.

▽ Try some for yourself

5. Draw a probability tree to show the total number of possible outcomes when tossing three coins.

Use your tree to find:

(a) the probability of 3 tails

(b) the probability of 2 heads and 1 tail.

6. An unbiased die is thrown twice

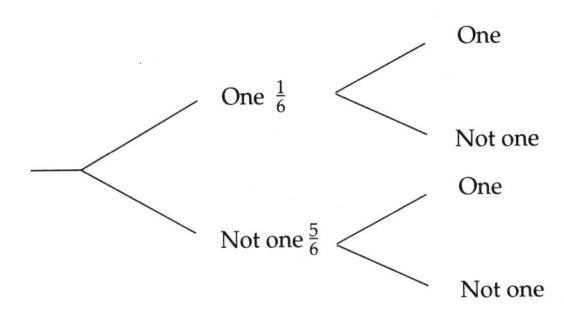

```
                   ┌── One
          One 1/6 ─┤
         /         └── Not one
────────┤
         \            ┌── One
          Not one 5/6 ┤
                      └── Not one
```

(a) Complete the diagram to show the probability of each outcome.

Use the diagram to find the probability that the score is:

(b) 2

(c) greater than 2.

We are asked to forecast the results of two hockey matches.
What is the probability of correctly forecasting both results?
What is the probability of correctly forecasting at least one result?

A match can have three results: win
                                  draw
                                  lose

Probability of a correct forecast = $\frac{1}{3}$

Probability of an incorrect forecast = $\frac{2}{3}$

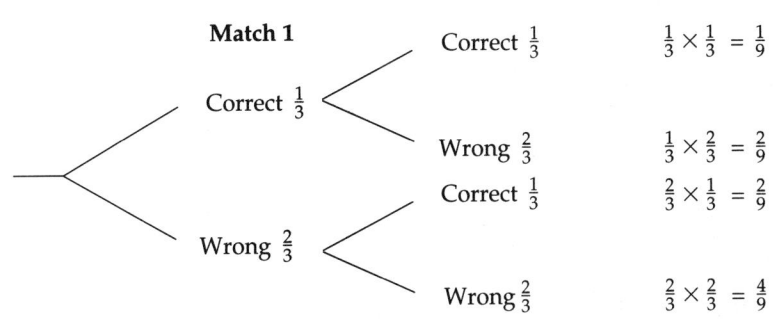

$$
\begin{array}{ccc}
& \textbf{Match 2} & \\
\textbf{Match 1} & & \\
& \text{Correct } \frac{1}{3} & \frac{1}{3} \times \frac{1}{3} = \frac{1}{9} \\
\text{Correct } \frac{1}{3} & & \\
& \text{Wrong } \frac{2}{3} & \frac{1}{3} \times \frac{2}{3} = \frac{2}{9} \\
& \text{Correct } \frac{1}{3} & \frac{2}{3} \times \frac{1}{3} = \frac{2}{9} \\
\text{Wrong } \frac{2}{3} & & \\
& \text{Wrong } \frac{2}{3} & \frac{2}{3} \times \frac{2}{3} = \frac{4}{9}
\end{array}
$$

Probability of both correct = $\frac{1}{9}$

Probability of at least one correct = $\frac{2}{9} + \frac{2}{9} + \frac{4}{9} = \frac{8}{9}$

What have you learnt about probability?

I have learnt that probability is not just about coins, dice and cards. Probability has lots of practical uses. Probability is not difficult if you are logical.

- [✓] I know how to work out simple probabilities as fractions.
- [✓] I can draw up a table to show all the possible outcomes of an event.
- [✓] I can construct a probability tree to show a sequence of events.
- [✓] I can use a probability tree by multiplying the probabilities from trunk to twig.
- [✓] I know that when I need to add probabilities from a tree I add from twig to twig.
- [✓] I know that probabilities add up to 1.

Further Questions

1. A group of 40 students on a GNVQ Built Environment course were going on an industrial visit. They would require safety boots so their shoes sizes were recorded in the table below.

Size	Number of pairs
8	9
9	15
10	10
11	3
12	3

Calculate the probability that a student requires:

(a) Size 9

(b) Size 11 or 12.

2. If the total amount invested in premium bonds is approximately £1 000 000 000 calculate the probability of winning a price with a £100 bond.

3. Draw up a table of all possible outcomes when 3 coins are tossed.

   Use your table to find the probability of:

   (a) 3 heads

   (b) 2 heads and 1 tail

   (c) 3 tails.

4. You are asked to use a suitable method to calculate the probability of a head followed by a 4 when you toss a coin and throw a die.

5. A GNVQ student is trying to find a part-time job in the catering industry. There are three possible jobs: at The Copper Kettle
   Pauline's Pantry
   Chez Charles.

   Six people apply for the job at The Copper Kettle. Five people apply for the job at Pauline's Pantry. Three people apply for the job at Chez Charles.

   The probability of winning the job at:
   The Copper Kettle is $\frac{1}{6}$
   Pauline's Pantry is $\frac{1}{5}$
   Chez Charles is $\frac{1}{3}$

   Draw a tree diagram showing all the possible probabilities.

   Use your diagram to calculate the probability that the student is offered:

   (a) All three jobs

   (b) A job at The Copper Kettle only

   (c) No job

   (d) One job only

6. A GNVQ Leisure and Tourism student forecasts the results of 3 football matches. Produce a probability three to show all the possible combinations.

   What is the probability that the student makes 3 accurate forecasts?

   What is the probability that at least 2 forecasts are correct?

Solutions

1. (a) Alphabet   26 letters   5 vowels

   Probability of a vowel $= \frac{5}{26}$

   (b) PROBABILITY   11 letters   4 vowels
   Probability of a vowel $= \frac{4}{11}$

   (c) APPLICATION OF NUMBER   19 letters
   8 vowels
   Probability of a vowel $= \frac{8}{19}$

2. Pack = 52 cards
   (a) 4 queens
   Probability of a queen = $\frac{4}{52} = \frac{1}{13}$
   (b) 2 black threes
   Probability of black three = $\frac{2}{52} = \frac{1}{26}$
   (c) 4 twos, 4 threes = 8
   Probability of 2 or 3 = $\frac{8}{52} = \frac{2}{13}$

3.

	1	2	3	4	5	6
1	2	3	4	5	6	7
2	3	4	5	6	7	8
3	4	5	6	7	8	9
4	5	6	7	8	9	10
5	6	7	8	9	10	11
6	7	8	9	10	11	12

(a) Probability 12 = $\frac{1}{36}$
(b) Probability 7 = $\frac{6}{36} = \frac{1}{6}$
(c) Probability of less than 7
    = $\frac{15}{36} = \frac{5}{12}$

4.

Coin 1	Coin 2	Coin 3	Probability
H	H	H	$\frac{1}{8}$
H	H	T	$\frac{1}{8}$
H	T	H	$\frac{1}{8}$
H	T	T	$\frac{1}{8}$
T	T	H	$\frac{1}{8}$
T	T	T	$\frac{1}{8}$
T	H	H	$\frac{1}{8}$
T	H	T	$\frac{1}{8}$

Probability of no heads = $\frac{1}{8}$
Probability of 1 head, 2 tails = $\frac{3}{8}$

5.

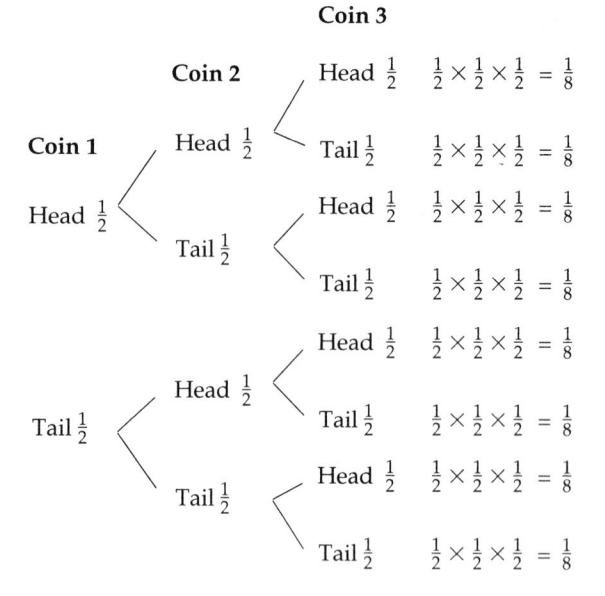

Coin 1 Head $\frac{1}{2}$
- Coin 2 Head $\frac{1}{2}$
  - Coin 3 Head $\frac{1}{2}$ $\frac{1}{2} \times \frac{1}{2} \times \frac{1}{2} = \frac{1}{8}$
  - Tail $\frac{1}{2}$ $\frac{1}{2} \times \frac{1}{2} \times \frac{1}{2} = \frac{1}{8}$
- Tail $\frac{1}{2}$
  - Head $\frac{1}{2}$ $\frac{1}{2} \times \frac{1}{2} \times \frac{1}{2} = \frac{1}{8}$
  - Tail $\frac{1}{2}$ $\frac{1}{2} \times \frac{1}{2} \times \frac{1}{2} = \frac{1}{8}$

Tail $\frac{1}{2}$
- Head $\frac{1}{2}$
  - Head $\frac{1}{2}$ $\frac{1}{2} \times \frac{1}{2} \times \frac{1}{2} = \frac{1}{8}$
  - Tail $\frac{1}{2}$ $\frac{1}{2} \times \frac{1}{2} \times \frac{1}{2} = \frac{1}{8}$
- Tail $\frac{1}{2}$
  - Head $\frac{1}{2}$ $\frac{1}{2} \times \frac{1}{2} \times \frac{1}{2} = \frac{1}{8}$
  - Tail $\frac{1}{2}$ $\frac{1}{2} \times \frac{1}{2} \times \frac{1}{2} = \frac{1}{8}$

(a) Probability 3 tails = $\frac{1}{8}$
(b) Probability 2 heads, 1 tail = $\frac{1}{8} + \frac{1}{8} + \frac{1}{8} = \frac{3}{8}$

6. (a)

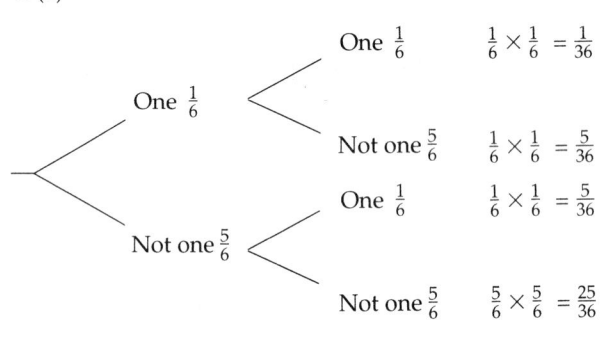

One $\frac{1}{6}$
- One $\frac{1}{6}$ $\frac{1}{6} \times \frac{1}{6} = \frac{1}{36}$
- Not one $\frac{5}{6}$ $\frac{1}{6} \times \frac{5}{6} = \frac{5}{36}$

Not one $\frac{5}{6}$
- One $\frac{1}{6}$ $\frac{5}{6} \times \frac{1}{6} = \frac{5}{36}$
- Not one $\frac{5}{6}$ $\frac{5}{6} \times \frac{5}{6} = \frac{25}{36}$

(b) Probability (score = 2) = Probability (1 + 1) = $\frac{1}{36}$
(c) Probability (score greater than 2) = $\frac{5}{36} + \frac{5}{36} + \frac{25}{36} = \frac{35}{36}$

12 Perimeters and Areas

Shapes, areas and volumes appear in some of my assignments. I think I know the names of most of the shapes, although some names are very strange. I shall need some help with areas and volumes.

Shapes, their areas and volumes appear in many practical situations. You should know the names of the most common shapes, and be able to find their areas. You will be expected to work out the volumes of some common solids.

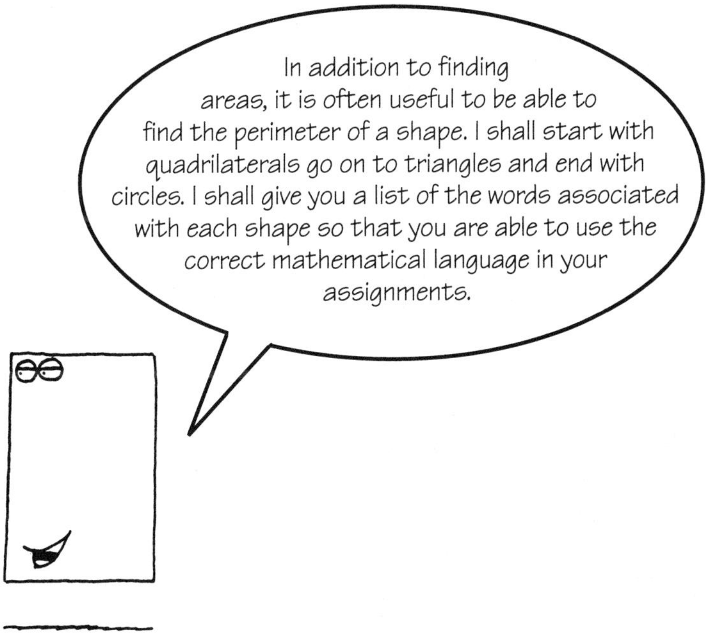

In addition to finding areas, it is often useful to be able to find the perimeter of a shape. I shall start with quadrilaterals go on to triangles and end with circles. I shall give you a list of the words associated with each shape so that you are able to use the correct mathematical language in your assignments.

179

Quadrilaterals

Words associated with quadrilaterals

Quadrilateral: any four-sided figure.

Angle: from the Latin word meaning a corner.

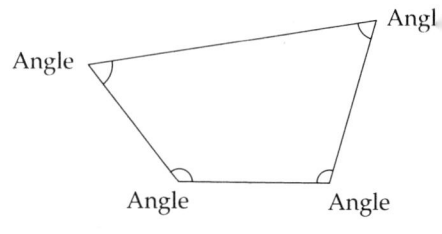

Right angle: an angle of 90°.

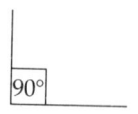

Square: a quadrilateral with all sides of the same length and all angles = 90°.

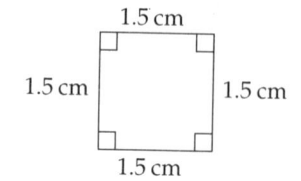

Rectangle: a quadrilateral with opposite sides equal and all angles = 90°.

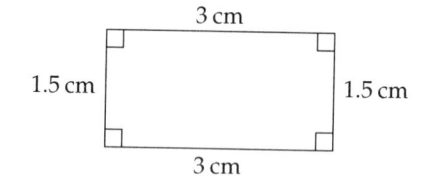

Parallelogram: a quadrilateral with two pairs of opposite sides equal and parallel.

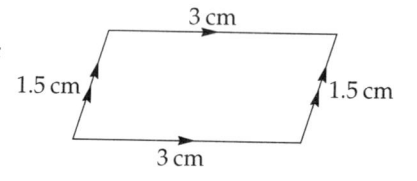

Trapezium: a quadrilateral with one pair of sides parallel.

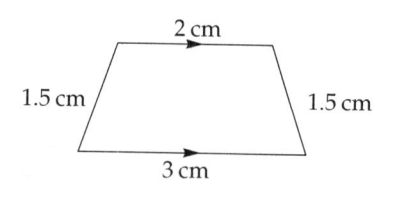

Parallel: two lines which are a constant distance apart.

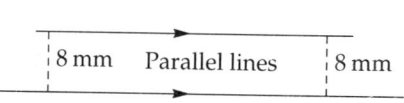

Diagonal: a line joining the opposite corners of a quadrilateral.

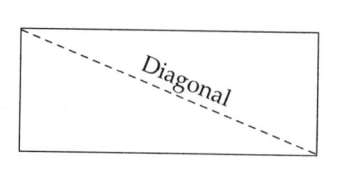

180

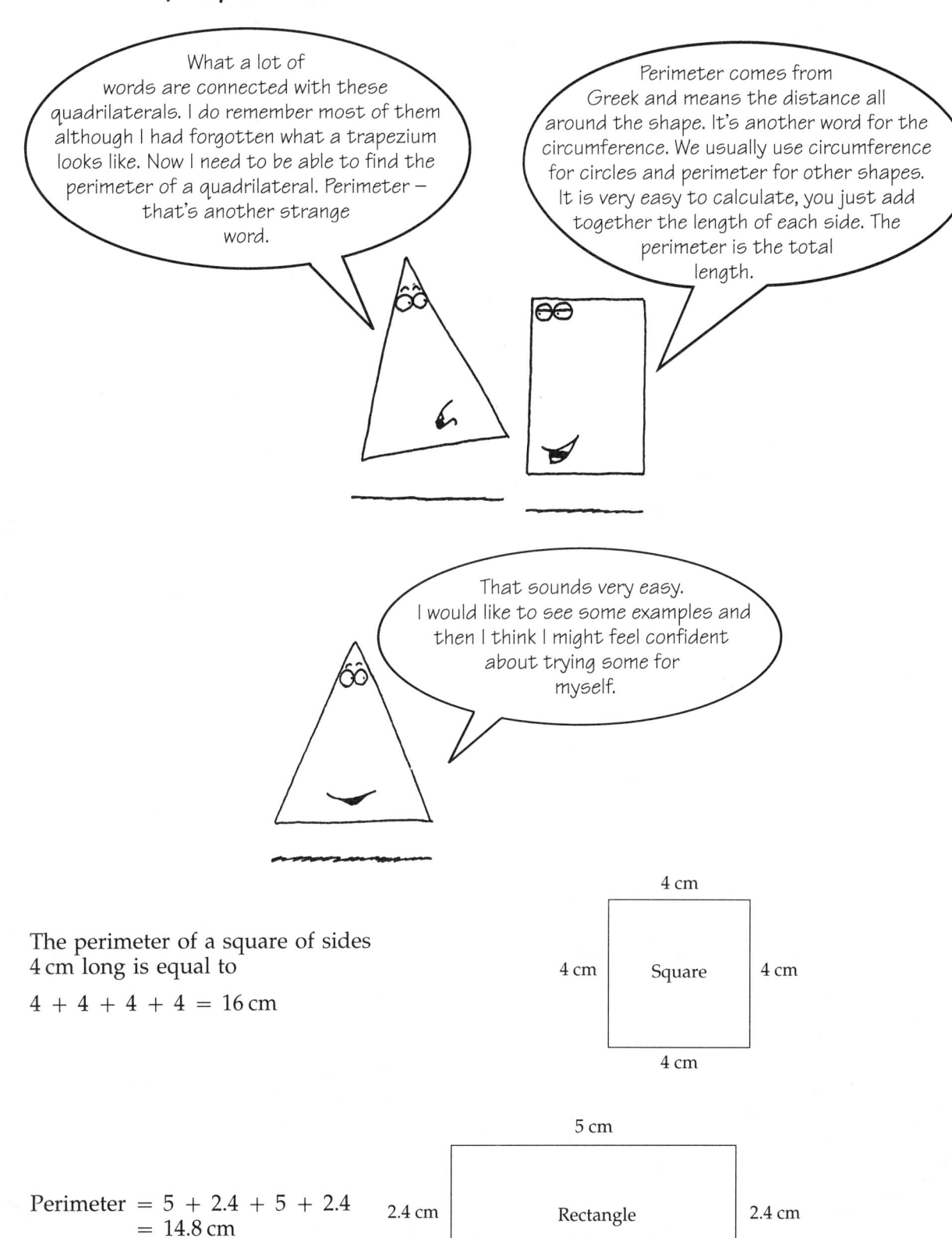

Perimeter of a quadrilateral

What a lot of words are connected with these quadrilaterals. I do remember most of them although I had forgotten what a trapezium looks like. Now I need to be able to find the perimeter of a quadrilateral. Perimeter — that's another strange word.

Perimeter comes from Greek and means the distance all around the shape. It's another word for the circumference. We usually use circumference for circles and perimeter for other shapes. It is very easy to calculate, you just add together the length of each side. The perimeter is the total length.

That sounds very easy. I would like to see some examples and then I think I might feel confident about trying some for myself.

The perimeter of a square of sides 4 cm long is equal to

4 + 4 + 4 + 4 = 16 cm

Perimeter = 5 + 2.4 + 5 + 2.4
= 14.8 cm

181

Perimeter = 5 + 2.4 + 5 + 2.4
= 14.8 cm

Parallelogram — 5 cm, 2.4 cm, 5 cm, 2.4 cm

Perimeter = 3.7 + 2.7 + 4.8 + 2.4
= 13.6 cm

Trapezium — 3.7 cm, 2.7 cm, 4.8 cm, 2.4 cm

Perimeter = 2.5 + 5 + 6.7 + 3.7
= 17.9 cm

Quadrilateral — 2.5 cm, 5 cm, 6.7 cm, 3.7 cm

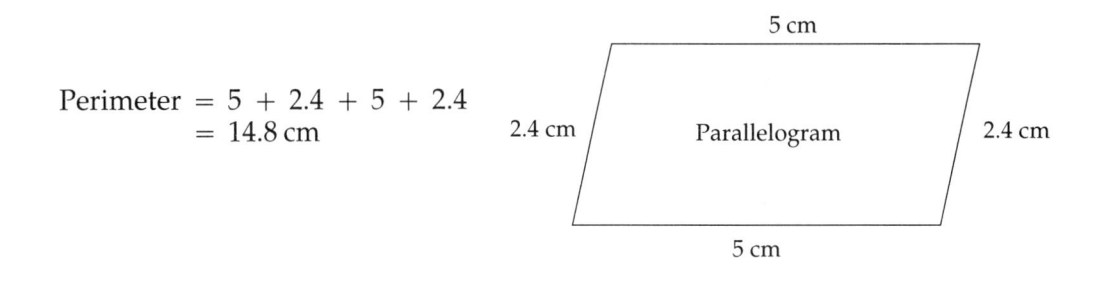

Try some for yourself

1. Find the perimeter of a quadrilateral with sides.
   (a) 5 cm, 5 cm, 5 cm, 5 cm
   (b) 12 cm, 3 cm, 12 cm, 3 cm
   (c) 12 cm, 4 cm, 10 cm, 5 cm
   (d) 4 cm, 5 cm, 6 cm, 3 cm

Area of some quadrilaterals

I had no trouble working out those perimeters. It's really quite straightforward. I am afraid I shall not find areas so easy. I think I can remember how to find some but I think I shall have trouble with the parallelogram and trapezium.

I shall go through each of the shapes in turn and show you how to work out the areas. The first thing we must think about is exactly what we mean by area, and how is it measured?

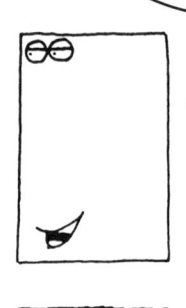

182

Area is the space inside the perimeter.
Area is always measured in square units which can be written as units^2.

A square centimetre is the area within a square of sides measuring 1 cm.

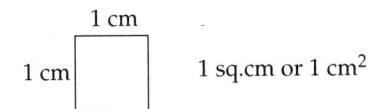

1 sq.cm or 1 cm^2

Now we shall look at a square of sides 3 cm.

We can mark in the number of 1 cm squares that will fit into this square.

There are 9 squares of area 1 cm^2

Total area = 9 × 1 cm^2 = 9 cm^2.

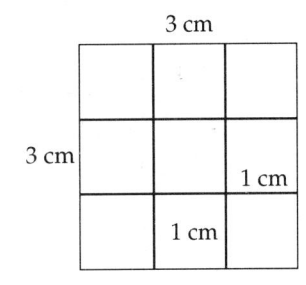

As this is not really a practical method of finding the area of a square, we need a formula.

Length of each side = l

Area = $l \times l$

Area of a square = l^2

Length of longer side = l

Length of shorter side = b

Area = $l \times b$

Area of rectangle = lb

You also need to know how to find the areas of a parallelogram and a trapezium. It is not so easy to show those by using unit squares. I will give you the formula for each so that you have it if you need to use it.

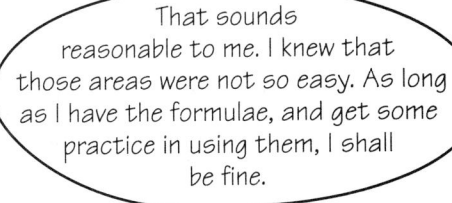

That sounds reasonable to me. I knew that those areas were not so easy. As long as I have the formulae, and get some practice in using them, I shall be fine.

> Both the parallelogram and the trapezium formulae require that you know the vertical height of the figure. You will always be given this, so it will just need to be substituted into the formula.

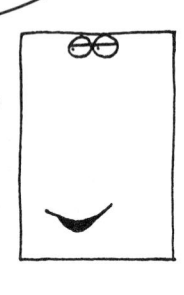

Area of parallelogram
$= 4 \times 1.6 \, \text{cm}^2$
$= 6.4 \, \text{cm}^2$

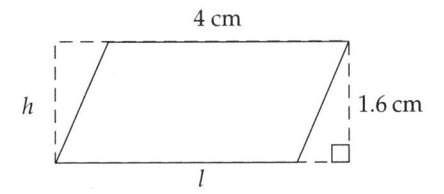

> Area of parallelogram = length × vertical height
> $= l \times h$
> $= lh$

Area of trapezium
$= \frac{1}{2}(3 + 6) \times 2.4 \, \text{cm}^2$
$= \frac{1}{2} \times 9 \times 2.4$
$= 10.8 \, \text{cm}^2$

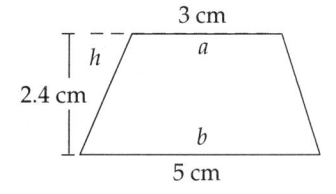

> Area of a trapezium = half the sum of the parallel sides × vertical height
> $= \frac{1}{2}(a + b) \times h$
> $= \frac{1}{2}(a + b)h$

Try some for yourself

Find the areas of the following shapes.

2.

(a) square, 2.5 cm × 2.5 cm

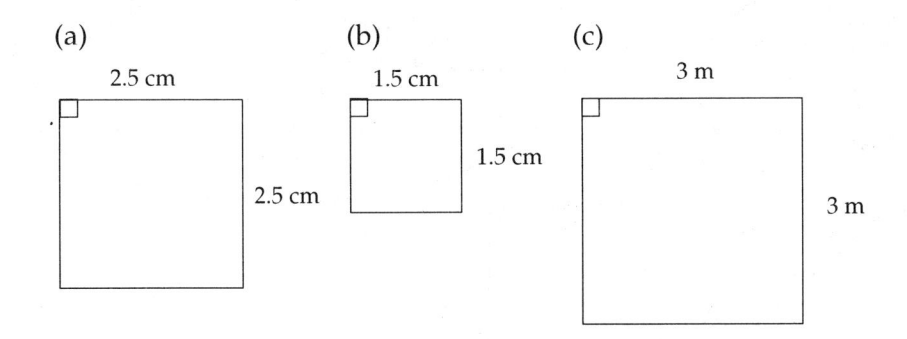

(b) square, 1.5 cm × 1.5 cm

(c) square, 3 m × 3 m

3.

(a) rectangle, 4 m × 2 m

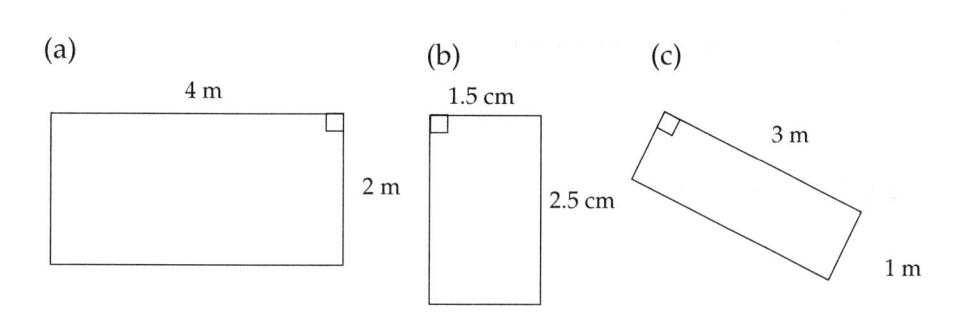

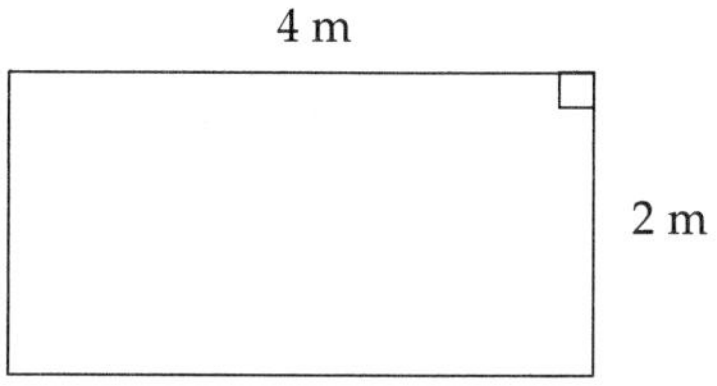

(b) rectangle, 1.5 cm × 2.5 cm

(c) rectangle, 3 m × 1 m

4.

(a) parallelogram, 3 m, 2.8 m

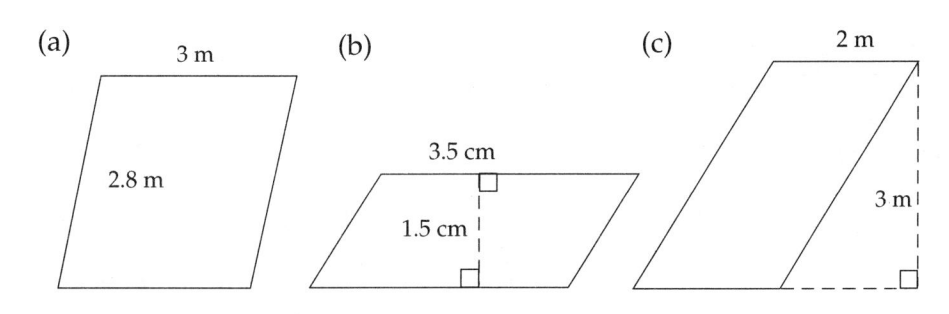

(b) parallelogram, base 3.5 cm, height 1.5 cm

(c) parallelogram, 2 m, height 3 m

5.

(a) trapezium, parallel sides 2 m and 4 m, height 2.5 m

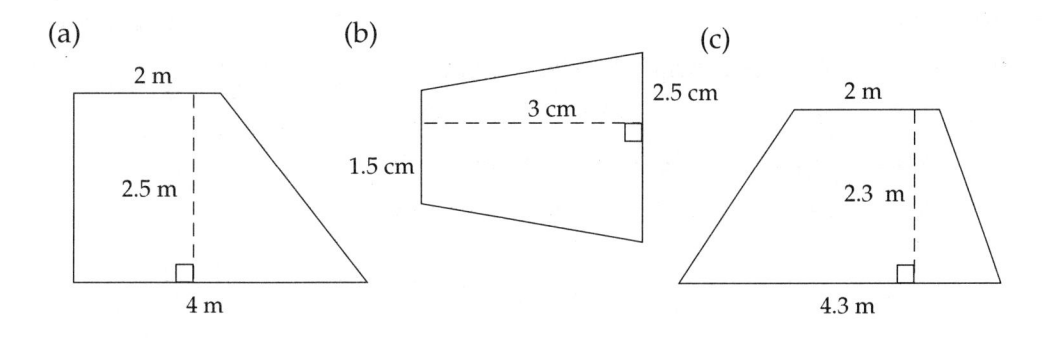

(b) trapezium, parallel sides 3 cm and 2.5 cm, height 1.5 cm

(c) trapezium, parallel sides 2 m and 4.3 m, height 2.3 m

185

Triangles

I need to know about triangles. I think I may be asked to find the perimeter and area of a triangle. The perimeter will just be the sum of the lengths of the sides I expect.

The perimeter of any figure is the sum of the lengths of the sides. The figure may have 3 sides or 30 sides, the perimeter is found in the same way – the 30-sided figure involves a much longer sum!

Perimeter of a triangle

Triangle ABC has sides
AB = 5 cm, AC = 6 cm, BC = 7 cm

Perimeter of ABC
= AB + AC + BC
= 5 + 6 + 7 = 18 cm

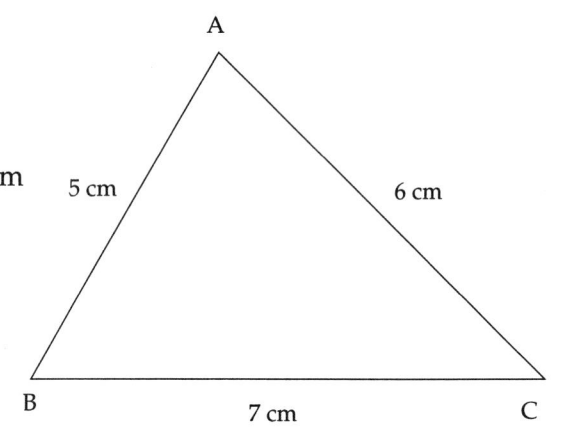

Try some for yourself

6. Find the perimeter of triangle ABC when:

(a) AB = 9.5 mm, AC = 10 mm, BC = 5 mm

(b) AB = 3 cm, AC = 4 cm, BC = 5 cm

(c) AB = 21 mm, AC = 21 mm, BC = 10 mm

These perimeters are so easy to work out. I expect the areas will be more difficult. Are we going to look at areas of triangles next?

No. Before we look at areas of triangles, I want to look at the angles of triangles. I also want to look at the 'special' triangles. They are the ones with the funny names – isosceles, equilateral.

Triangle: The word triangle means three-angled, so the angles are very important.

Angles of a triangle: The angles are measured in degrees and *always* total 180°.

> Sum of the angles of a triangle is 180°

If I know two angles in a triangle I can always find the third.

Angle A + Angle B + Angle C = 180°

58° + ?° + 70° = 180°

58° + 70° = 128°

Angle C = 180° − 128° = 52°

Check: 58° + 52° + 70° = 180°

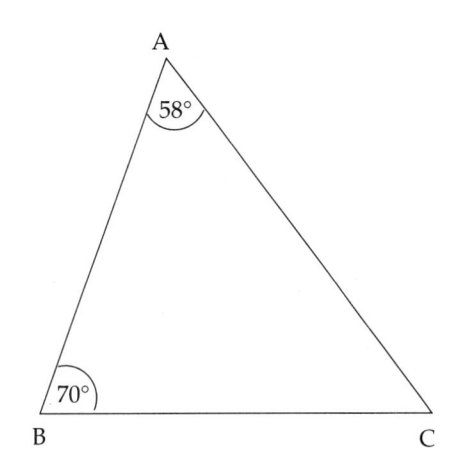

Angle	A	B	C
Value	?°	90°	20°

Angle B + Angle C
= 90° + 20° = 110°

Angle A = 180° − 110° = 70°

Check: 70° + 90° + 20° = 180°

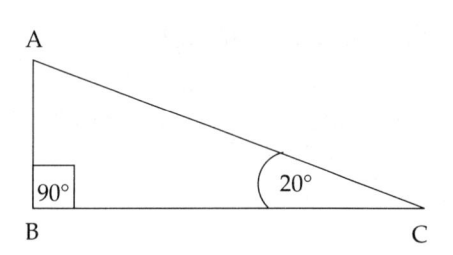

187

▽ Try some for yourself

7. Find the missing angles and complete the table.

	Angle A	Angle B	Angle C
(a)	110°	37°	
(b)	90°		41°
(c)	60°	60°	
(d)		55°	55°

Types of triangle

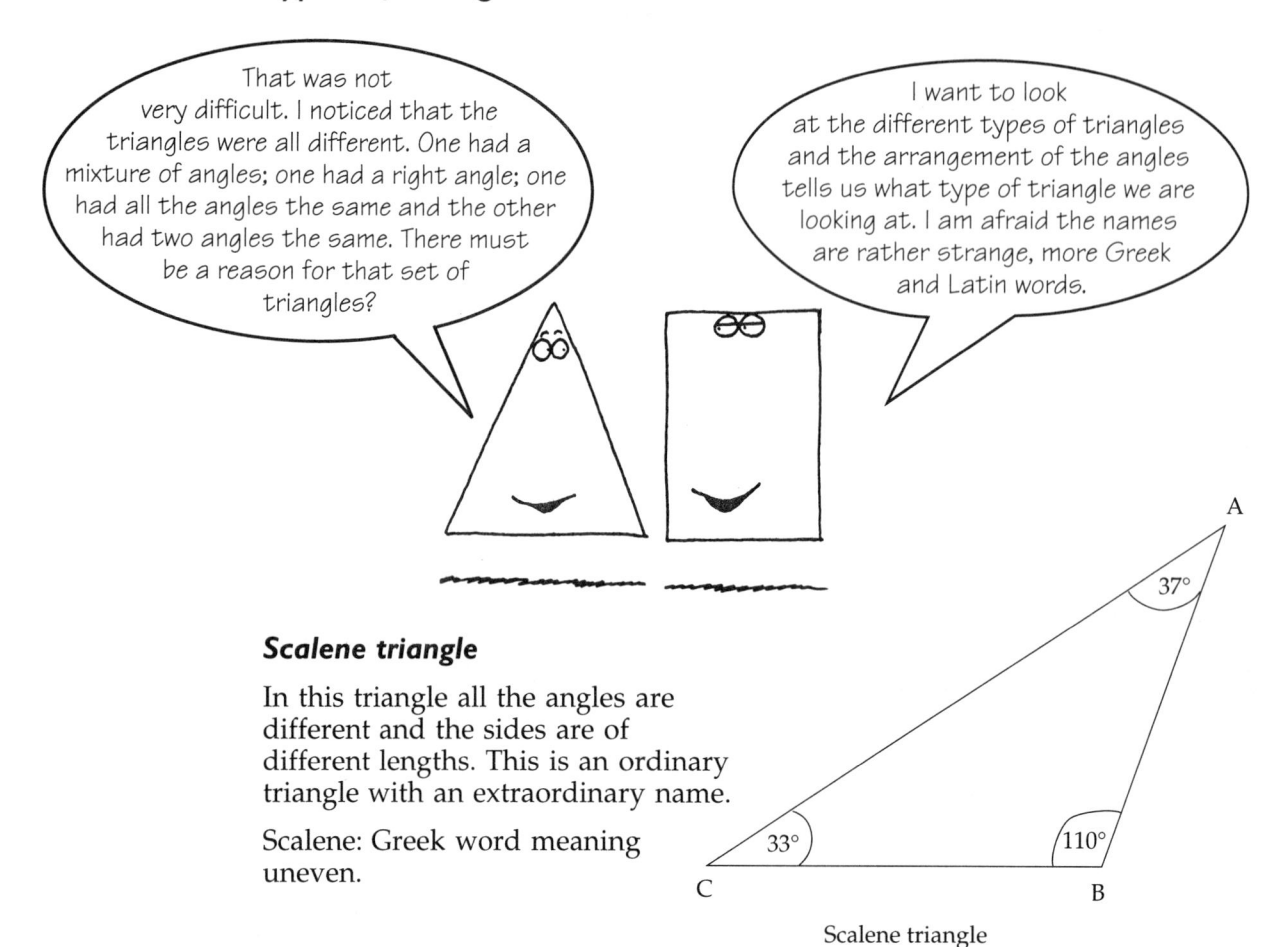

That was not very difficult. I noticed that the triangles were all different. One had a mixture of angles; one had a right angle; one had all the angles the same and the other had two angles the same. There must be a reason for that set of triangles?

I want to look at the different types of triangles and the arrangement of the angles tells us what type of triangle we are looking at. I am afraid the names are rather strange, more Greek and Latin words.

Scalene triangle

In this triangle all the angles are different and the sides are of different lengths. This is an ordinary triangle with an extraordinary name.

Scalene: Greek word meaning uneven.

Scalene triangle

Right-angled triangle

This triangle has one angle, at A, measuring 90°. That means that this is a very special triangle – a right-angled triangle.

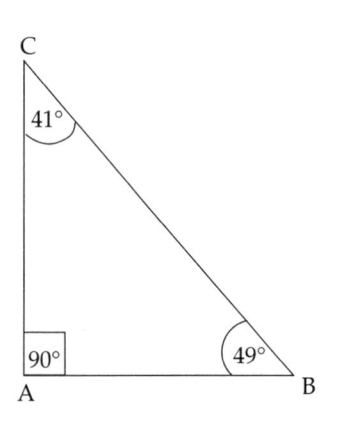

Right-angled triangle

188

Equilateral triangle

In this triangle all the angles are equal and measure 60°. If we measure the sides we find that all the sides are the same length.

Equilateral: Latin meaning equal sides.

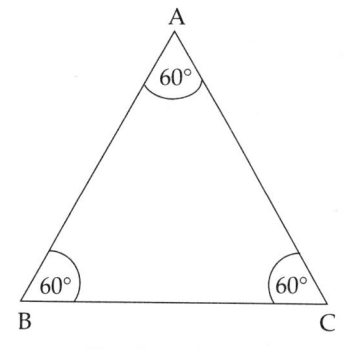

Equilateral triangle

Isosceles triangle

In this triangle two angles are the same size and the two sides leading to the angles are of equal length.

Isosceles: Greek meaning equal legs.

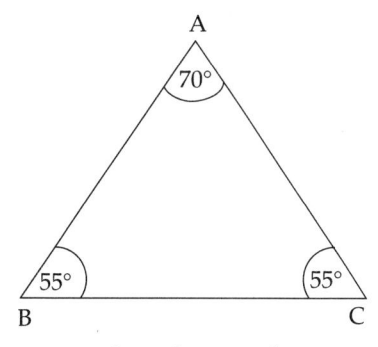

Isosceles triangle

▽ Try some for yourself

8. Identify the type of triangle.

	Angle A	Angle B	Angle C
(a)	75°	86°	19°
(b)	60°	60°	60°
(c)	35°	55°	90°
(d)	36°	108°	36°

Area of a triangle

Is all this information about triangles really necessary? I'm sure I will never need to use this knowledge when I leave this course. It all seems so pointless.

You may be surprised to find that this knowledge is useful later – it depends on your job. I was explaining about the types of triangle because I need to use that knowledge to find the area of a triangle. I want you to tell me what you can remember about rectangles – everything.

189

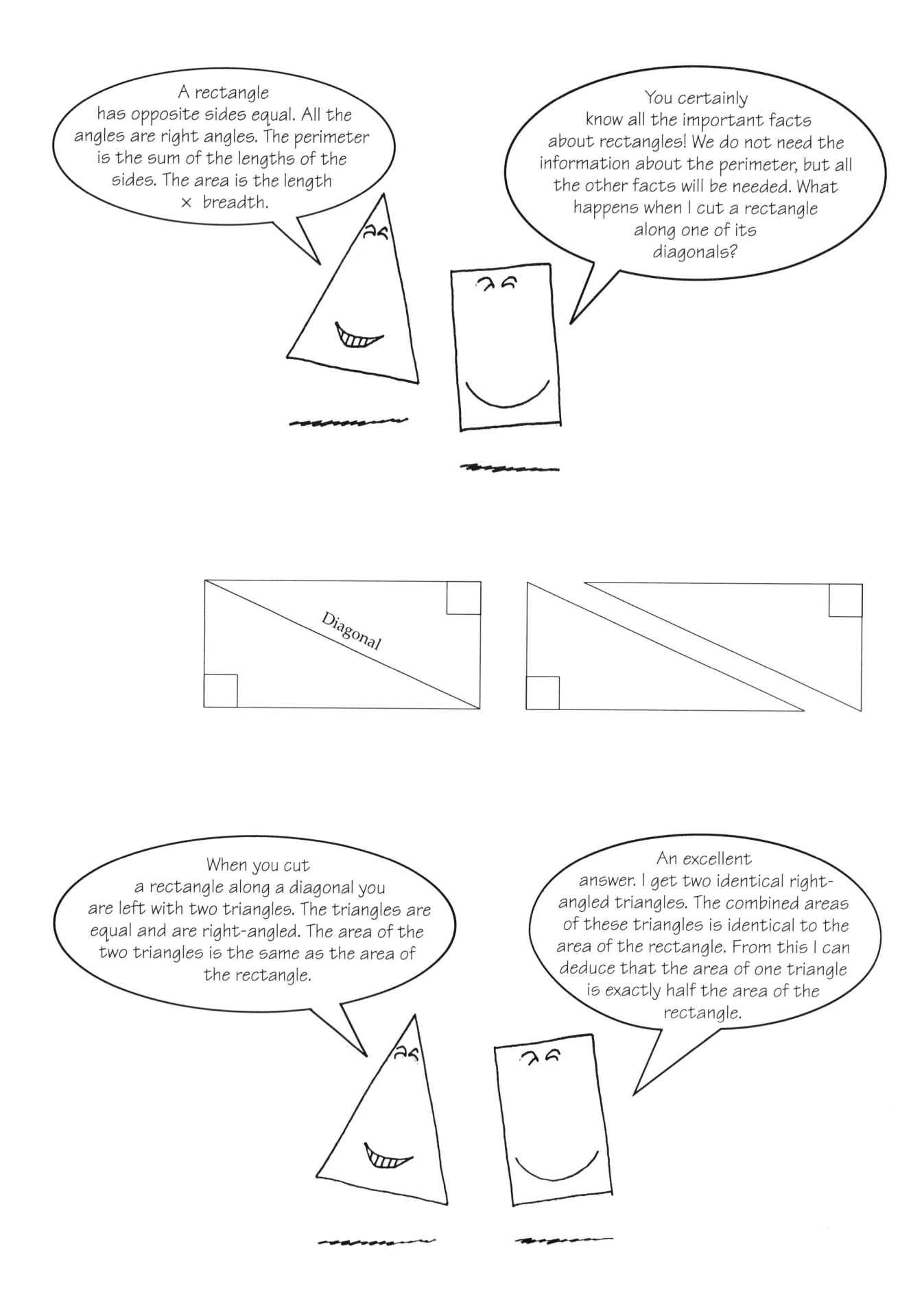

A rectangle has opposite sides equal. All the angles are right angles. The perimeter is the sum of the lengths of the sides. The area is the length × breadth.

You certainly know all the important facts about rectangles! We do not need the information about the perimeter, but all the other facts will be needed. What happens when I cut a rectangle along one of its diagonals?

Diagonal

When you cut a rectangle along a diagonal you are left with two triangles. The triangles are equal and are right-angled. The area of the two triangles is the same as the area of the rectangle.

An excellent answer. I get two identical right-angled triangles. The combined areas of these triangles is identical to the area of the rectangle. From this I can deduce that the area of one triangle is exactly half the area of the rectangle.

Area of triangle = $\frac{1}{2} lb$

Just to confuse you, we use different words for length and breadth when looking at triangles. The length is called the base; the breadth is called the height.

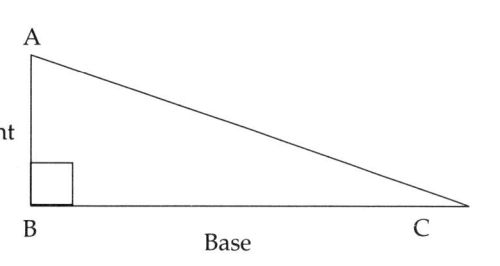

Area of triangle = $\frac{1}{2} bh$

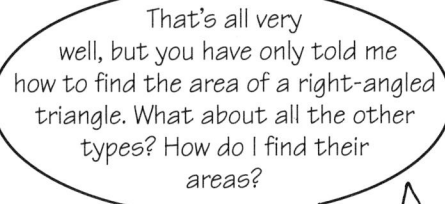

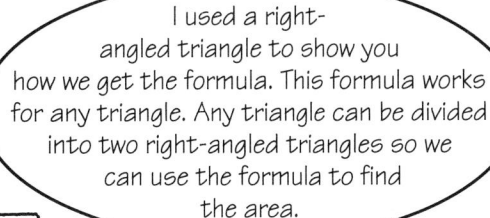

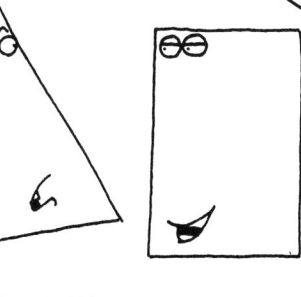

That's all very well, but you have only told me how to find the area of a right-angled triangle. What about all the other types? How do I find their areas?

I used a right-angled triangle to show you how we get the formula. This formula works for any triangle. Any triangle can be divided into two right-angled triangles so we can use the formula to find the area.

Area of triangle ABC
= Area of triangle ABD + area of triangle ADC

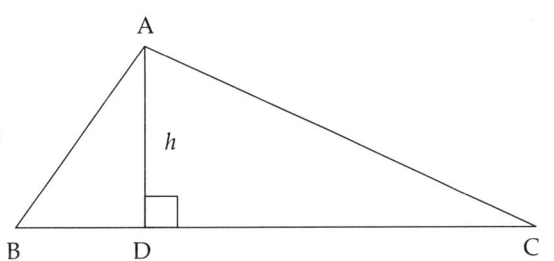

Area of ABD = $\frac{1}{2}$ BD × AD
Area of ADC = $\frac{1}{2}$ DC × AD
Area of ABC = $\frac{1}{2}$ BD × AD + $\frac{1}{2}$ DC × AD = $\frac{1}{2}$ AD (BD + DC)
   BD + DC = BC
   Area of ABC = $\frac{1}{2}$ BC × AD
             = $\frac{1}{2}$ base × height

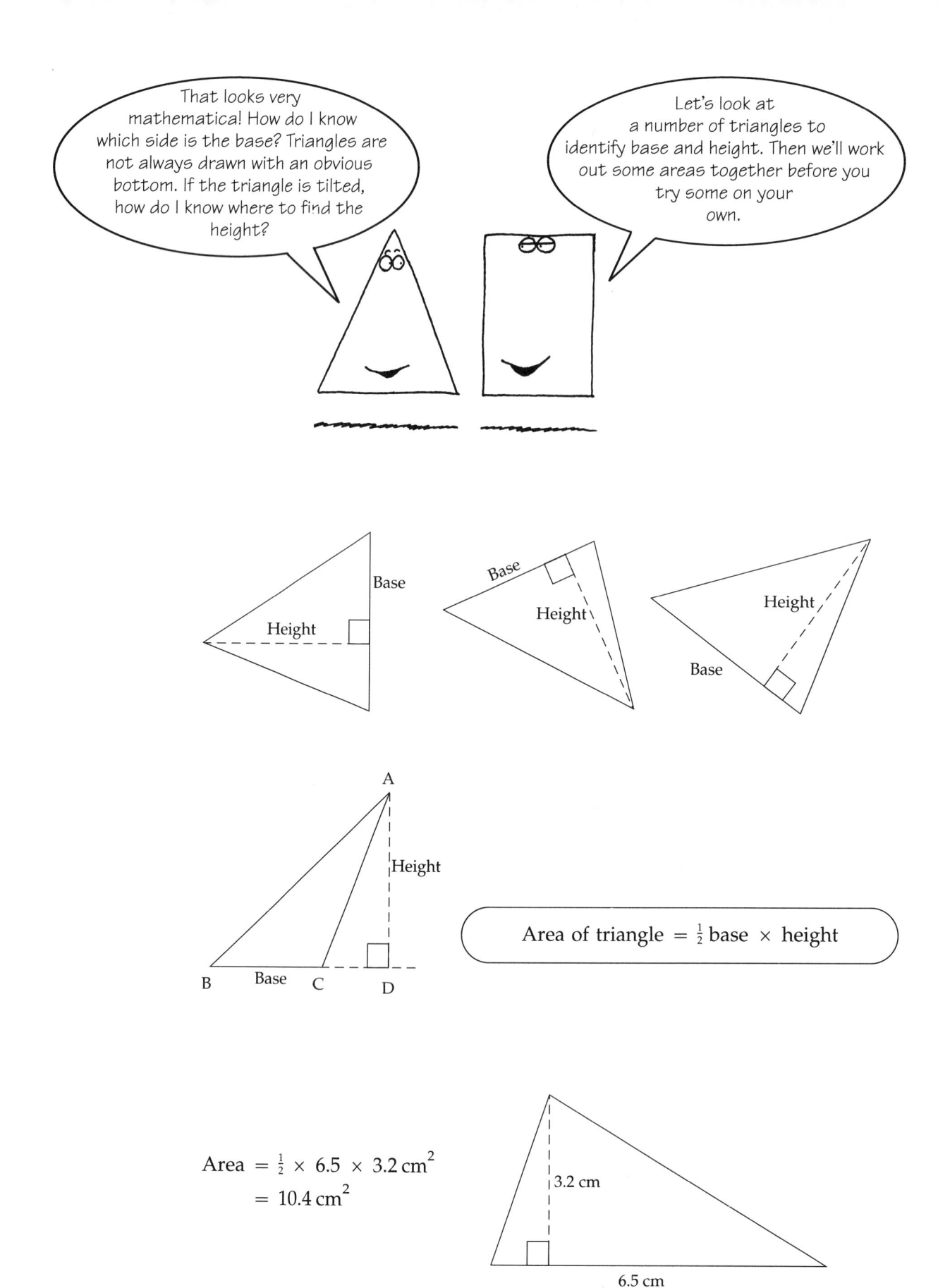

That looks very mathematical! How do I know which side is the base? Triangles are not always drawn with an obvious bottom. If the triangle is tilted, how do I know where to find the height?

Let's look at a number of triangles to identify base and height. Then we'll work out some areas together before you try some on your own.

Area of triangle = $\frac{1}{2}$ base × height

Area = $\frac{1}{2}$ × 6.5 × 3.2 cm^2
    = 10.4 cm^2

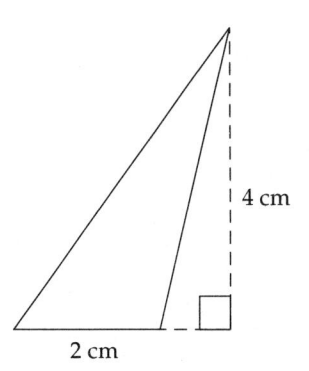

Area = $\frac{1}{2}$ × 2 × 4 cm^2
     = 4 cm^2

4 cm

2 cm

Try some for yourself

9. Find the area of these triangles.

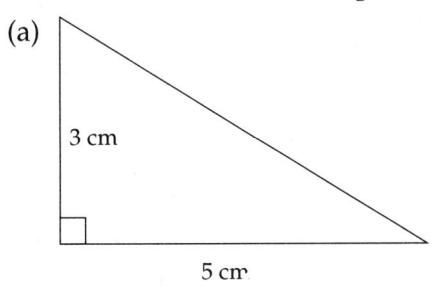

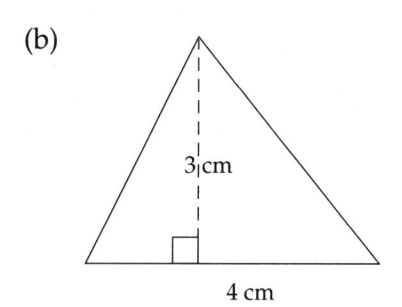

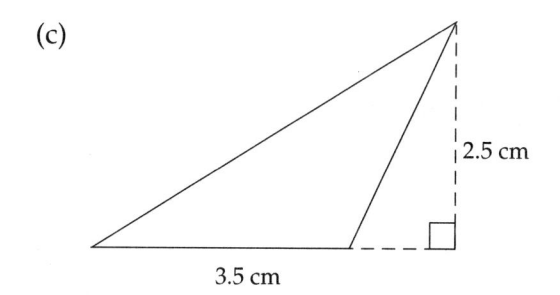

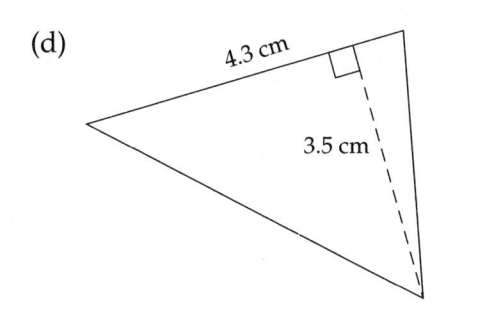

(a)

3 cm

5 cm

(b)

3 cm

4 cm

(c)

2.5 cm

3.5 cm

(d)

4.3 cm

3.5 cm

193

Circles

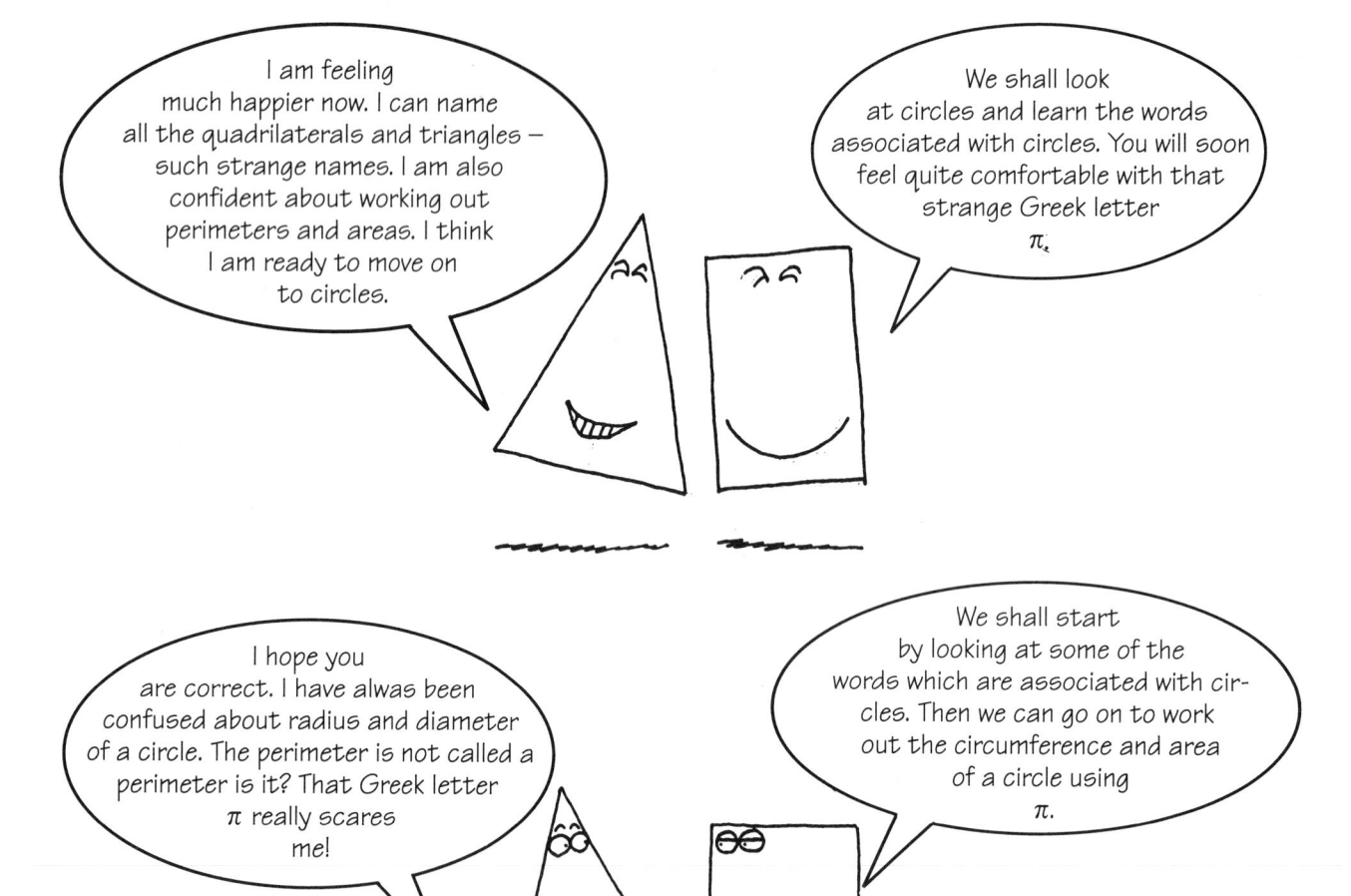

I am feeling much happier now. I can name all the quadrilaterals and triangles — such strange names. I am also confident about working out perimeters and areas. I think I am ready to move on to circles.

We shall look at circles and learn the words associated with circles. You will soon feel quite comfortable with that strange Greek letter π.

I hope you are correct. I have alwas been confused about radius and diameter of a circle. The perimeter is not called a perimeter is it? That Greek letter π really scares me!

We shall start by looking at some of the words which are associated with circles. Then we can go on to work out the circumference and area of a circle using π.

Words associated with circles

Centre: the point at the middle of the circle.

Circumference: the distance all round the circle.

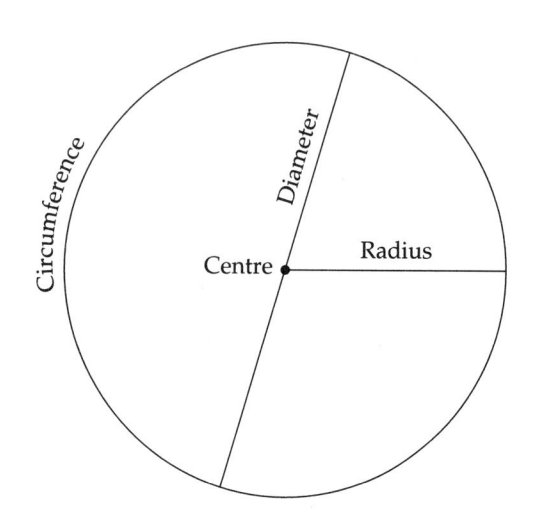

194

8.1 cm

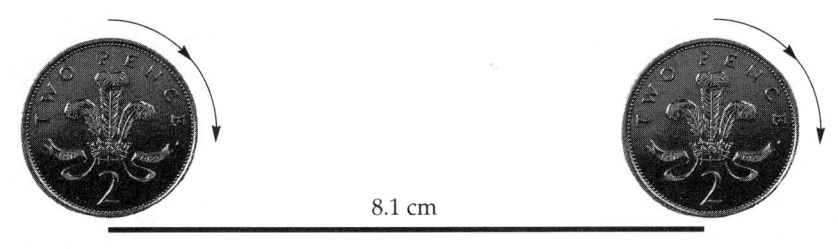

The circumference is the distance rolled by a circle before coming the right way up again. The circumference of the 2p piece is shown by the length of the line.

Radius: the distance from the centre of the circle to any point on the circumference.

Diameter: any straight line joining two points on the circumference and passing through the centre.

Circumference of a circle

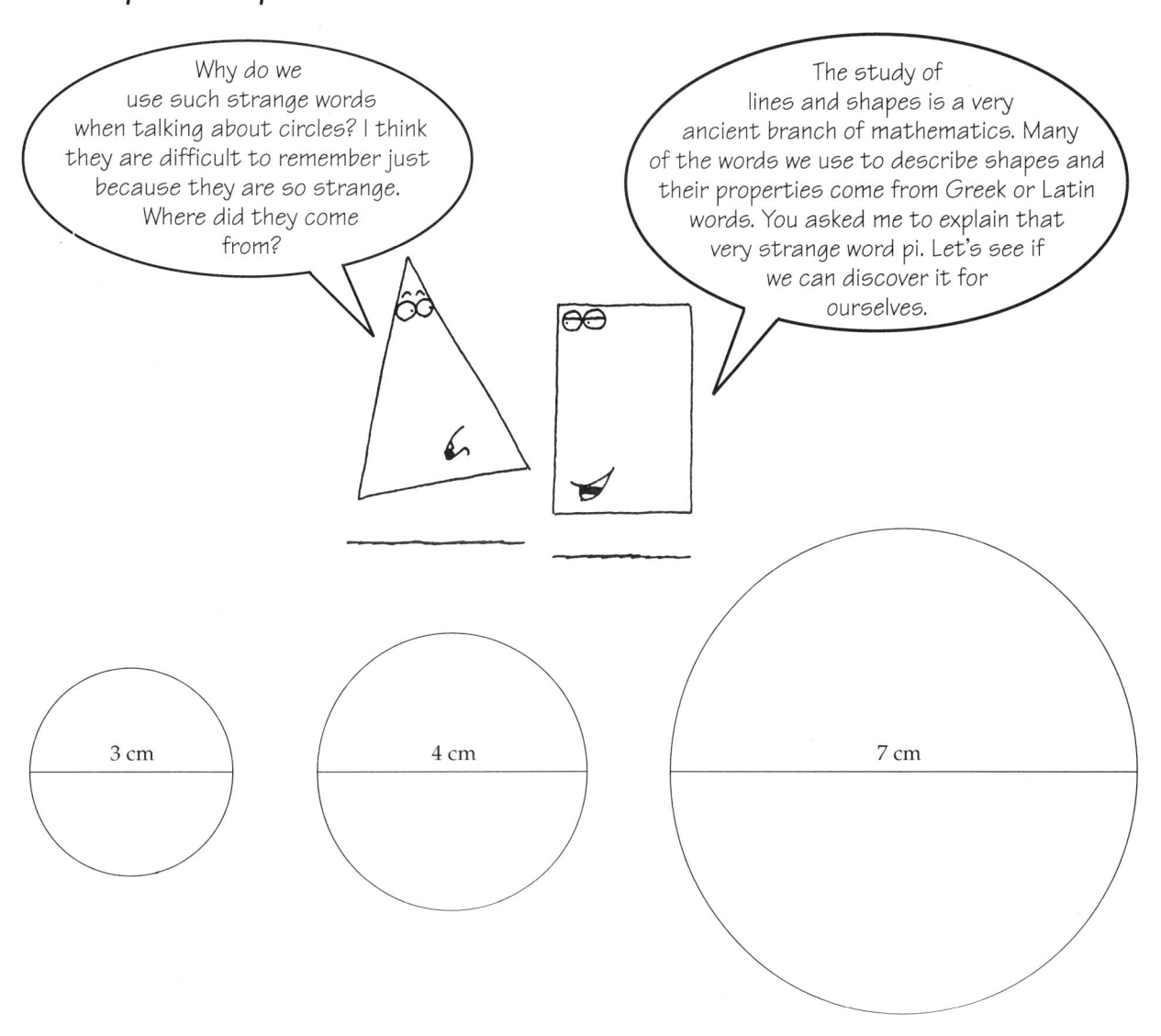

Why do we use such strange words when talking about circles? I think they are difficult to remember just because they are so strange. Where did they come from?

The study of lines and shapes is a very ancient branch of mathematics. Many of the words we use to describe shapes and their properties come from Greek or Latin words. You asked me to explain that very strange word pi. Let's see if we can discover it for ourselves.

3 cm

4 cm

7 cm

What can you tell me about these circles? I want you to look at the diameter and circumference of each circle. It is a simple question with a simple commonsense answer.

The simplest answer is that as the diameter increases, the circumference also increases. The circle with a diameter of 3 cm has a smaller circumference than the circle with a diameter of 4 cm. The circle with a diameter of 7 cm has a large circumference. Is that what you wanted?

That is exactly what I was looking for. The circumference of a circle increases as the diameter increases. Now I want to carry out a practical investigation involving diameters, circumferences and ending with pi!

Here are four objects found around the house. There is a mug, a small vase, a basin and a litter bin. The circumference and diameter of each were measured. The results are recorded in the table.

Object	Diameter	Circumference
mug	7.5 cm	24.5 cm
vase	7 cm	22 cm
basin	17 cm	53 cm
bin	24.5 cm	77 cm

That's really interesting! I bet it was fun taking all those measurements. What am I supposed to learn from these figures and how do measurements of diameter and circumference tell me anything about pi?

That is only the first part of the investigation. I have to divide each circumference by its diameter. I think the results will surprise you.

Object	Circumference ÷ Diameter	Result to 2 d.p.
mug	24.5 ÷ 7.5	3.27
vase	22 ÷ 7	3.14
basin	53 ÷ 17	3.15
bin	77 ÷ 24.5	3.14

That's pretty neat! The results were all very close. The measurements were all different so how can the results be so similar? What does all this have to do with pi?

As you said, all the results were very similar. If the measurements had been really accurate all the results would have given me a value of 3.14. The result should be a constant value. This constant value is called pi and is usually written using the Greek letter π.

I never realised what pi meant before. We were just told to use it but no one ever explained where it came from. I shall remember that now, it means something at last.

π is used in formulae connected with circles. The formula for the circumference of a circle includes π.

C stands for circumference
d stands for diameter
r stands for radius

$$C = \pi \times d = \pi d$$

or $C = \pi \times 2 \times r = 2\pi r$

If the diameter of a circle is 20 cm then the radius is 10 cm.

Remember: the diameter is twice as long as the radius

$$C = 20\pi = 62.83 \text{ to 2 d.p.}$$
$$C = 2\pi \times 10 = 62.83 \text{ to 2 d.p.}$$

Try some for yourself

π is usually taken to be 3.142
It can also be written as the fraction $\frac{22}{7}$.

A scientific calculator has a π button. You can use that if it is easier for you.

Give all answers to 2 d.p.

10. Find the circumference of a circle with:

 (a) radius 21 cm

 (b) radius 49 cm

 (c) radius 4 cm

11. Find the circumference of a circle with:

 (a) diameter 42 cm

 (b) diameter 98 cm

 (c) diameter 8 cm

Those were easy. What happens if I know the circumference but need to find the radius or the diameter? Do I use the same formulae?

You use the same formulae but they need to be rearranged to make it easier for you. I'll rearrange them for you in case you need to find a radius or a diameter.

$C = \pi d$ can be rearranged to give you the diameter

$$d = \frac{C}{\pi} \qquad d = \frac{62.83}{\pi} = 20 \text{ cm}$$

$C = 2\pi r$ can be rearranged to give you the radius.

$$r = \frac{C}{2\pi} \qquad r = \frac{62.83}{2\pi} = 10 \text{ cm}$$

Try some for yourself

All answers should be given to 2 decimal places.

12. Find the diameter of a circle with circumference:

(a) 132 cm

(b) 33 cm

(c) 62 cm

13. Find the radius of a circle with circumference:

(a) 75 cm

(b) 86 cm

(c) 62 cm

Area of a circle

I think I am quite clear about finding the circumference of a circle. I can also find the radius and diameter if I only know the circumference. What about the area? I seem to remember that π comes into that formula too.

That's correct. π is very important when we work with circles. It comes into the formula for the area of a circle. I'm pleased that you have remembered that. I think I will just remind you of the formula for the area of a circle. You may not remember it exactly.

199

$$\text{Area} = \pi \times r \times r = \pi r^2$$

Find the area of a circle of radius 2 cm

$$A = \pi 2^2 = 4\pi = 12.57 \text{ cm}^2 \text{ to 2 d.p.}$$

▽ Try some for yourself

14. Find the area of a circle of radius:

(a) 21 cm

(b) 49 cm

(c) 4 cm

15. Find the area of a circle of diameter:

(a) 42 cm

(b) 98 cm

(c) 8 cm

Remember: the radius is equal to the diameter divided by 2.

You may also need to rearrange this formula for the area of a circle. Sometimes you may know the area but need to find the radius. This is the formula you will need.

$$r^2 = \frac{A}{\pi} \quad \text{so } r = \sqrt{\frac{A}{\pi}}$$

▽ Try some for yourself

16. Find the radius of a circle of area:

(a) 1885 cm^2

(b) 27 cm^2

(c) 50 cm^2

Composite shapes

All this work on areas of quadrilaterals, triangles and circles seems very easy. There must be something more complicated that I shall be asked to do with areas.

As usual, you are correct. So far we have looked at a number of basic shapes and discovered how to work out the areas. In many practical situations you may be asked to find the area of a shape that is made from more than one basic shape. These are called **composite shapes**.

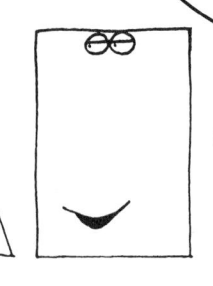

This figure can be divided into two rectangles. This will allow us to find the area of the L-shape.

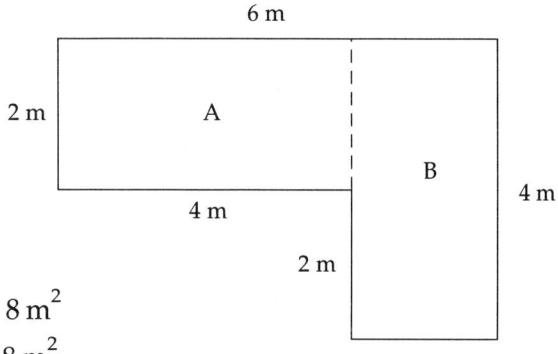

Rectangle A: Area = 4 × 2 m^2 = 8 m^2
Rectangle B: Area = 4 × 2 m^2 = 8 m^2
Area of L = Area of A + Area of B = 8 + 8 m^2
= 16 m^2

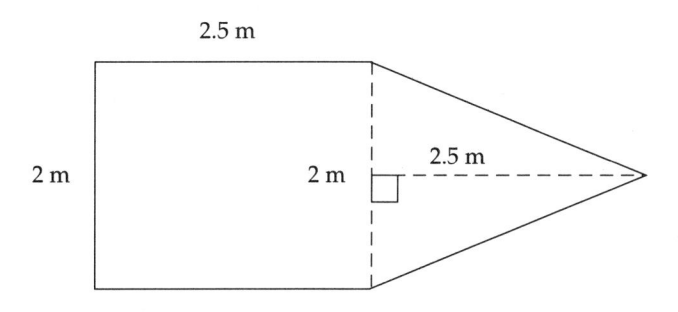

Area of triangle = $\frac{1}{2}$ × 2 × 2.5 m^2
= 2.5 m^2
Area of rectangle = 2 × 2.5 m^2
= 5 m^2
Total area = area of triangle + area of rectangle = 2.5 + 5 = 7.5 m^2

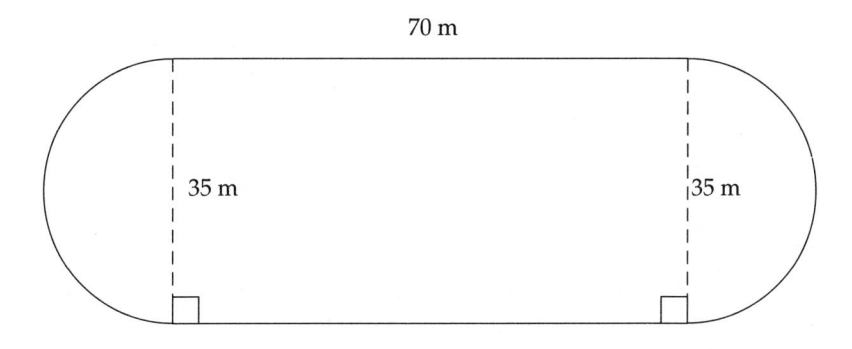

This running track can be divided into a rectangle and two semicircles.

Area of rectangle = 70 × 35 m^2
= 2450 m^2

Area of a circle = πr^2

Area of a semicircle = $\dfrac{\pi r^2}{2}$

Area of semicircular part = $\dfrac{\pi \times 17.5^2}{2}$ m^2
= 481 m^2

Area of 2 semicircles of radius 17.5 m = 2 × 481 m^2
= 962 m^2

Area of running track = 2450 + 962 = 3412 m^2

▽ Try some for yourself

Find the areas of these composite shapes.

17. (a)

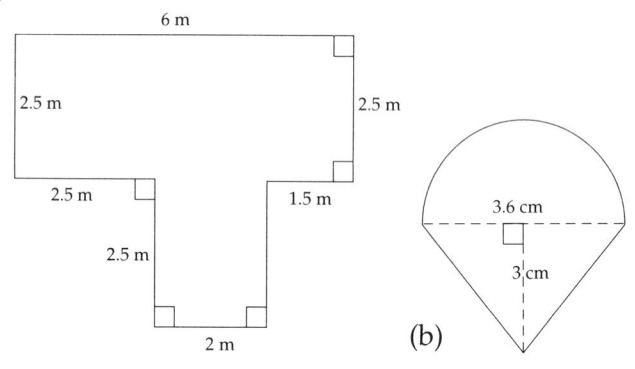

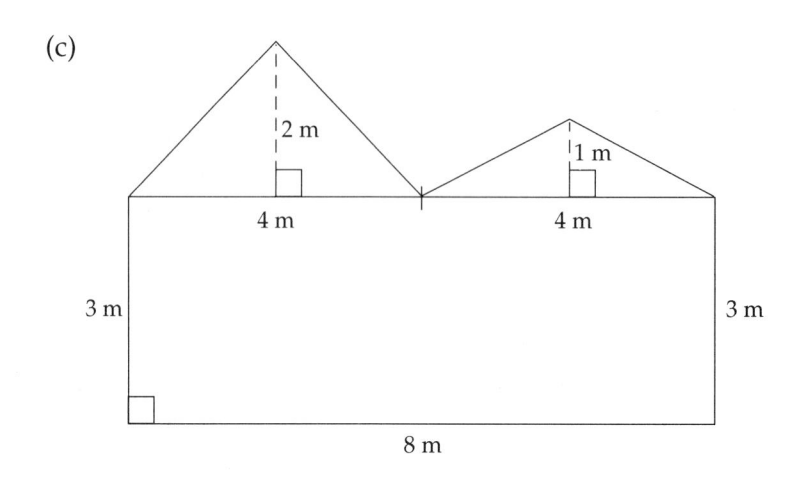

(b)

(c)

What have you learnt about perimeters and areas?

I was rather confused about perimeters and areas. I did not understand why I needed to use π. Now I am much more confident and feel I understand much more.

- ✓ I know that the perimeter is the distance all round the shape.
- ✓ I know that the perimeter of a circle is called the circumference.
- ✓ I know why we use π in calculations involving circles.
- ✓ I know the names of quadrilaterals and triangles.
- ✓ I know that area is the space inside the perimeter of a shape.
- ✓ I know that area is always measured in square units.
- ✓ I know that the angles in a triangle always add up to 180°.
- ✓ I know how to find the areas of triangles, quadrilaterals and circles.

Further Questions

1. A room has dimensions as shown in the diagram.

   New carpet is to be laid. It comes in a standard size of 2.7 m × 3.6 m.

   (a) Find the total area to be carpetted.

   (b) Find the area of carpet that will be wasted.

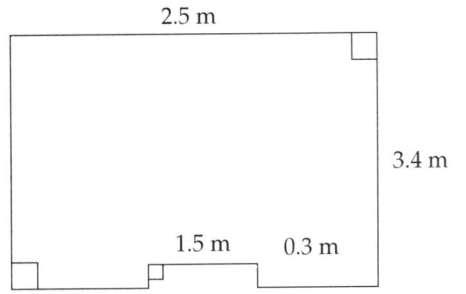

   Dimensions: 2.5 m, 3.4 m, 1.5 m, 0.3 m

2. An office measures 2.5 m square. It contains a number of pieces of furniture.

       1   4 drawer cabinet 0.45 m × 0.6 m
       2   desks each 1.5 m × 0.75 m
       1   bookcase 0.75 m × 0.3 m
       1   table 1 m × 0.6 m

   (a) Calculate the area of carpet needed to cover the floor.

   (b) Calculate the area of floor space occupied by the furniture.

   (c) Calculate the area of floor space which is not occupied by furniture.

3. A patio, measuring 2 m × 3 m is to be paved. The paving slabs are 0.6 m × 0.6 m.

   How many slabs will be needed to pave the patio?

4. The diagram shows the plan of a netball court. All dimensions are marked.

   Find:

   (a) the perimeter of the court

   (b) the area of the court

   (c) the area of the centre third

   (d) the area of the centre circle

   (e) the area of the goal circle

   (f) the circumference of the centre circle.

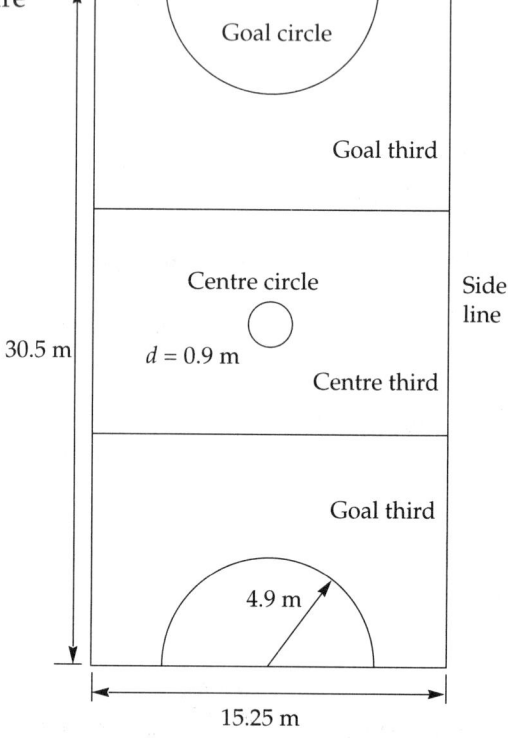

5. A football pitch is 120 m by 92 m. Find the perimeter and area of the pitch.

6. The sizes of paper are given.

       A5   105 × 148.5 mm
       A4   210 × 297 mm
       A3   297 × 420 mm

   Find: (a) the perimeter of each size.

          (b) the area of each size.

7. Wrapping paper is sold in two sizes.

   (a) 90 cm × 260 cm

   (b) 75 cm × 4 m

   Find the area of paper for each size.

8. The plan of a hospital ward is given. All dimensions are marked.

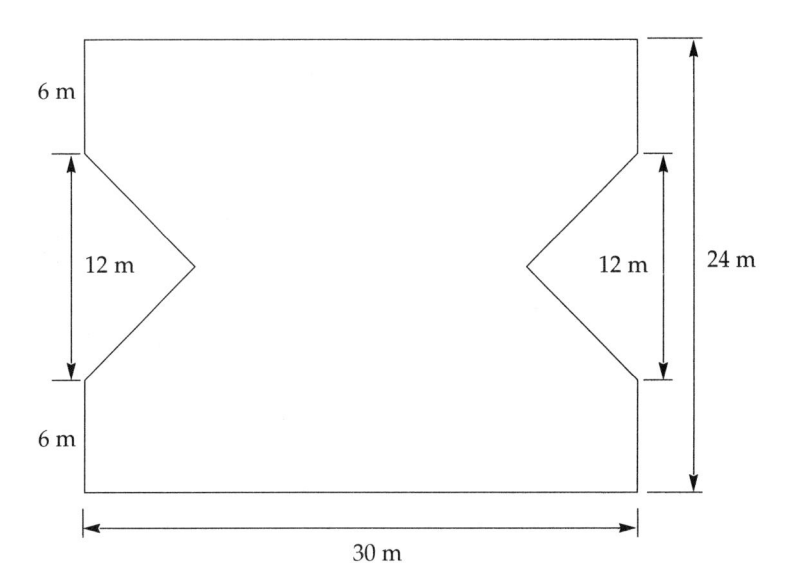

   Each bed and locker requires 2.5 m of wall space.

   (a) How many beds can be fitted into the ward?

   (b) How much floor area is available in the ward?

9.

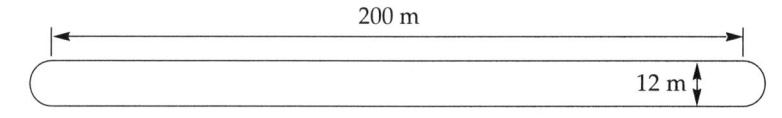

   A running track has two semicircular ends of diameter 12 m. There are two long straights of 200 m.

   (a) Find the perimeter of the track.

   (b) Find the area of the track.

10. A restaurant has a large circular table of radius 1 m (= 100 cm). The waiter is told to seat customers 65 cm apart.

    (a) How many customers can he seat?

    (b) If the distance between customers is reduced to 62 cm, how many will he be able to seat?

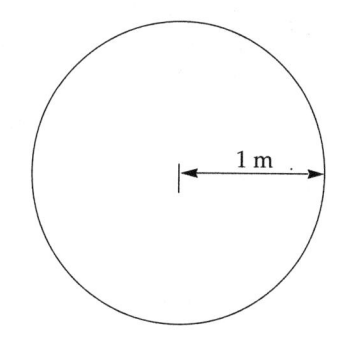

11. A room measures 3 m by 4 m. Carpet tiles measuring 0.25 m by 0.25 m are to be fitted.

(a) How many tiles will be required?

(b) Tiles are sold in boxes of five. How many boxes will be needed?

12. Find the lengths of the bus routes shown below.

(a) 2.5 km, 2 km, 1.5 km, 2 km, 2.5 km

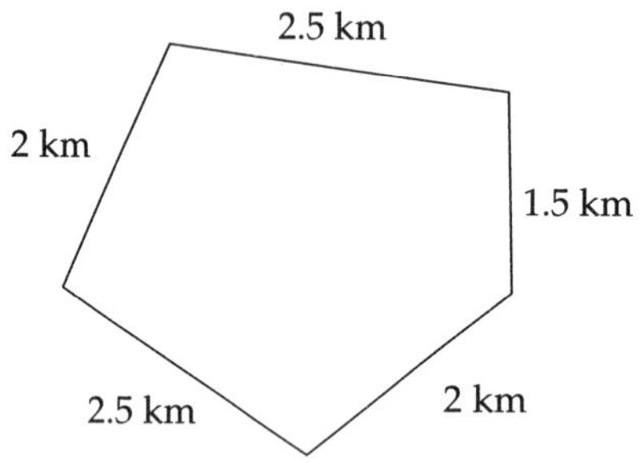

(b) 1.5 km, 1.5 km, 2 km, 4.5 km, 2.5 km

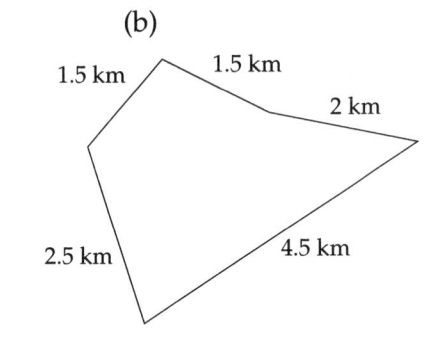

(c) 2 km, 5 km, 3.5 km, 4 km, 5 km, 3 km, 4 km

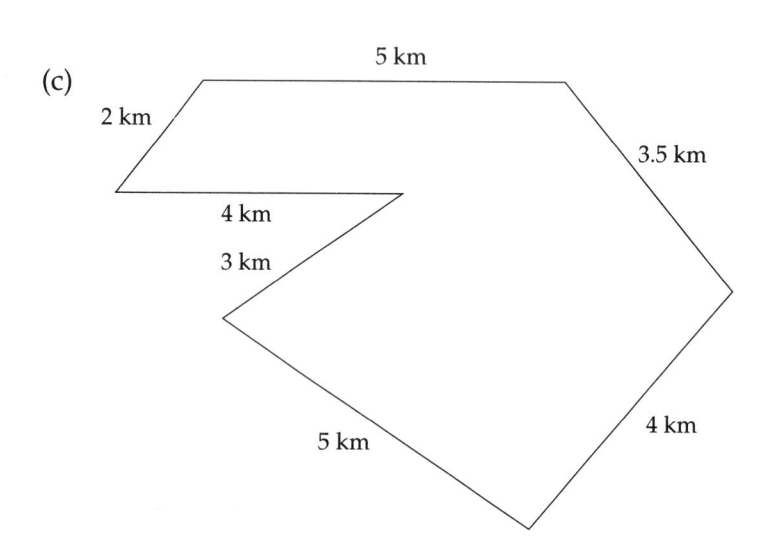

13. A dartboard is in the shape of a circle with radius 22 cm. The scoring area is within a circle of radius 17 cm. The centre circle has a radius of 1.8 cm.

(a) Find the area of the dartboard.

(b) Find the scoring area.

(c) Find the area of the centre circle.

(d) Find the non-scoring area.

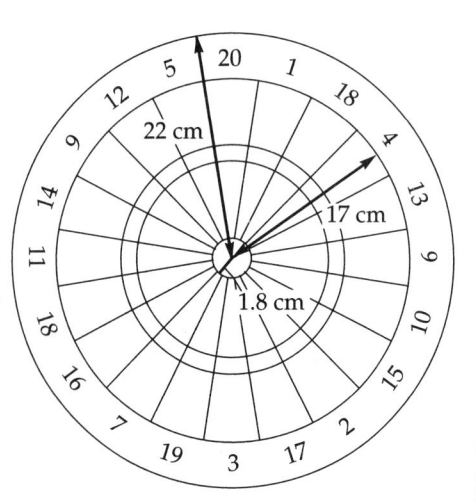

14. A baking sheet measuring 32 cm by 22 cm is to be given a non-stick coating. Find the area to be coated.

15. A basketball court measures 26 m by 14 m. The distance between the boundary of the court and the spectators must be 1 m.

(a) Calculate the area of the court.

(b) Work out the perimeter of the court.

(c) Find the total length of rope needed to provide a barrier for spectators all around the court.

16. The diagram shows the side of a swimming pool. The depth is 1 m at the shallow end and 6 m at the deep end.

Find the area of the side.

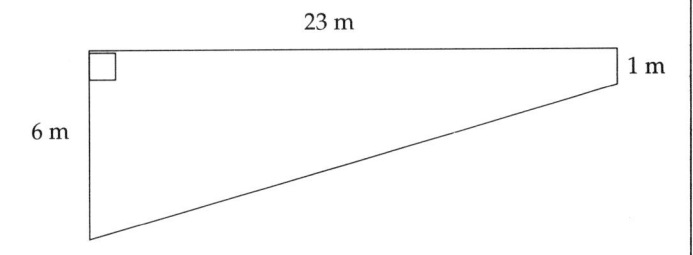

17. The side wall of a squash court is in the shape of a trapezium.

Find the area of the side wall.

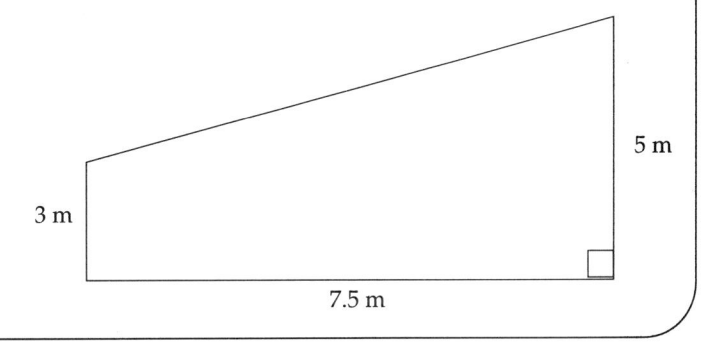

Solutions

1. (a) 5 + 5 + 5 + 5 = 20 cm
(b) 12 + 3 + 12 + 3 = 30 cm
(c) 12 + 4 + 10 + 5 = 31 cm
(d) 4 + 5 + 6 + 3 = 18 cm

2. (a) 2.5 × 2.5 = 6.25 m²
(b) 1.5 × 1.5 = 2.25 cm²
(c) 3 × 3 = 9 m²

3. (a) 4 × 2 = 8 m²
(b) 2.5 × 1.5 = 3.75 cm²
(c) 3 × 1 = 3 m²

4. (a) 3 × 2.8 = 8.4 m²
(b) 3.5 × 1.5 = 5.25 cm²
(c) 2 × 3 = 6 m²

5. (a) $\frac{1}{2}$ (2 + 4) × 2.5 = 7.5 m²
(b) $\frac{1}{2}$ (1.5 + 2.5) × 3 = 6 cm²
(c) $\frac{1}{2}$ (2 + 4.3) × 2.3 = 7.245 m²

6. (a) 9.5 + 10 + 5 = 24.5 mm
(b) 3 + 4 + 5 = 12 cm
(c) 21 + 21 + 10 = 52 mm

7.

	Angle A	Angle B	Angle C
(a)	110°	37°	**33°**
(b)	90°	**49°**	41°
(c)	60°	60°	**60°**
(d)	**70°**	55°	55°

8. (a) Scalene

(b) Equilateral

(c) Right-angled

(d) Isosceles

9. (a) $\frac{1}{2} \times 5 \times 3 = 7.5 \text{ cm}^2$

(b) $\frac{1}{2} \times 4 \times 3 = 6 \text{ cm}^2$

(c) $\frac{1}{2} \times 3.5 \times 2.5 = 4.375 \text{ cm}^2$

(d) $\frac{1}{2} \times 4.3 \times 3.5 = 7.525 \text{ cm}^2$

10. (a) 131.95 cm to 2 d.p.

(b) 307.88 cm to 2 d.p.

(c) 25.13 cm to 2 d.p.

11. (a) 131.95 cm to 2 d.p.

(b) 307.88 cm to 2 d.p.

(c) 25.13 cm to 2 d.p.

12. (a) 42.02 cm to 2 d.p.

(b) 10.50 cm to 2 d.p.

(c) 19.74 cm to 2 d.p.

13. (a) 11.94 cm to 2 d.p.

(b) 13.69 cm to 2 d.p.

(c) 9.87 cm to 2 d.p.

14. (a) 1385.33 cm² to 2 d.p.

(b) 7542.96 cm² to 2 d.p.

(c) 50.26 cm² to 2 d.p.

15. (a) 1385.44 cm² to 2 d.p.

(b) 7542.96 cm² to 2 d.p.

(c) 50.26 cm² to 2 d.p.

16. (a) 21 cm to 2 d.p.

(b) 2.93 cm to 2 d.p.

(c) 3.99 cm to 2 d.p.

17. (a) Area of rectangle $6 \times 2.5 \text{ m} = 15 \text{ m}^2$
Area of rectangle $2 \times 2.5 \text{ m} = 5 \text{ m}^2$
Total area $= 20 \text{ m}^2$

(b) Semicircle diameter 3.6 cm, radius $3.6 \div 2 \text{ cm} = 1.8 \text{ cm}$

Area of semicircle $= \frac{1}{2}\pi \times 1.8^2$
$= 5.1 \text{ cm}^2$ to 2 d.p.

Area of triangle $= \frac{1}{2} \times 3.6 \times 3 = 5.4 \text{ cm}^2$

Total area $= 10.5 \text{ cm}^2$ to 2 d.p.

(c) Area of rectangle $3 \times 8 \text{ m} = 24 \text{ m}^2$

Area of triangle $\frac{1}{2} \times 4 \times 2 \text{ m} = 4 \text{ m}^2$

Area of small triangle $\frac{1}{2} \times 4 \times 1 \text{ m} = 2 \text{ m}^2$

Total area $= 30 \text{ m}^2$

13 Volumes

> Now I am confident about finding perimeters and areas I need to move on to volumes. I vaguely remember that the formulae for volumes are connected with the formulae for areas.

> The volume of a solid is simply the area of its cross-section multiplied by the length of the solid. I think it will be useful to look at solids in the same order that I followed for areas. I'll look first at cubes and cuboids, then at a triangular prism, finishing with one solid connected with circles: the cylinder.

> That seems quite a logical way to deal with volumes. Cubes and cuboids must be connected with quadrilaterals. Why do we have two words – don't they mean the same thing? They sound so similar.

209

Cube

A cube is a three-dimensional shape. The length, width and height are all equal. It looks like a solid square.

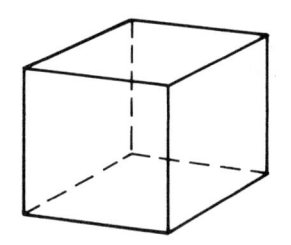

Cuboid

A cuboid is a three-dimensional shape. The end, or cross-sectional area, is rectangular in shape. A cuboid has height as well as length and width. It looks like a solid rectangle.

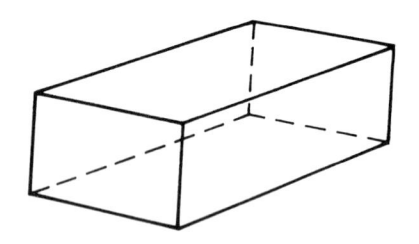

What is volume?

Volume is the measure of the space filled by a solid object like a building block. Volume is measured in cubic measures.

A cubic unit is the space inside a cube whose sides measure 1 unit.

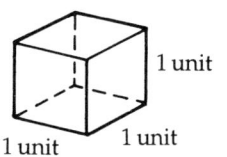

Cubes and cuboids

This cube, measuring 2 units by 2 units by 2 units contains 8 cubic units.

$2 \times 2 \times 2 = 8$ cubic units

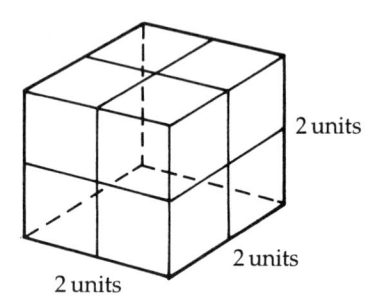

A cube with length 2 cm, width 2 cm and height 2 cm has a volume of $2\,cm \times 2\,cm \times 2\,cm = 8$ cubic cm or $8\,cm^3$

$$\text{Volume of a cube} = \text{length} \times \text{width} \times \text{height}$$

$$V = l\,w\,h$$

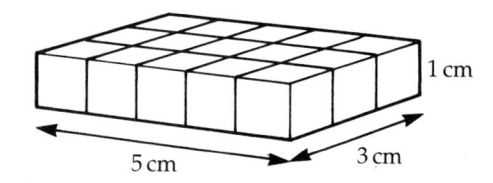

A cuboid is a solid rectangle.

210

End (or cross-sectional) area = length × width
= 5 × 1 cm^2

Volume = cross-sectional area × height
= 5 × 1 cm^3
= 15 cm^3

> Volume of a cuboid = length × width × height

> $V = l\,w\,h$

Triangular prisms

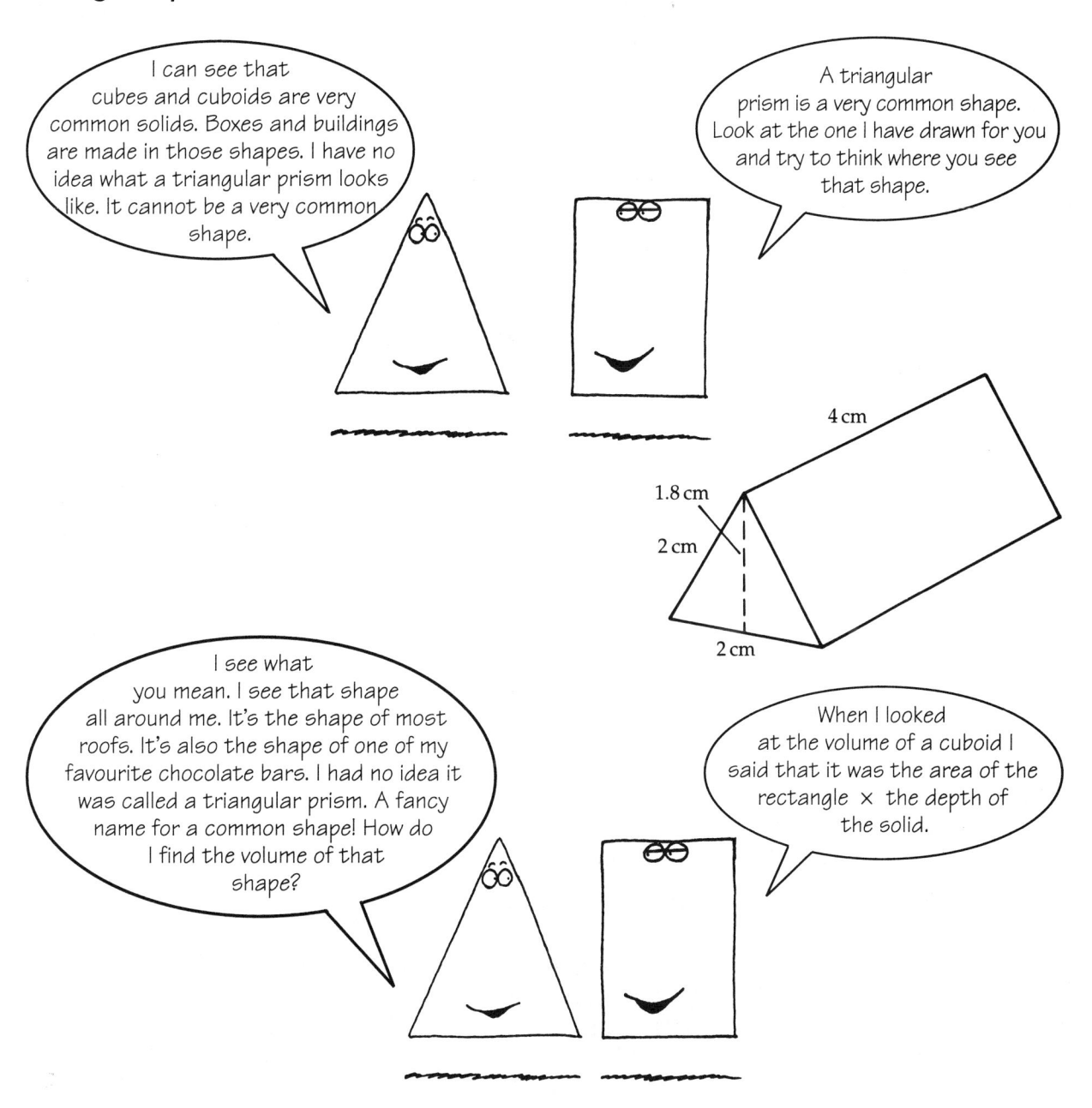

I can see that cubes and cuboids are very common solids. Boxes and buildings are made in those shapes. I have no idea what a triangular prism looks like. It cannot be a very common shape.

A triangular prism is a very common shape. Look at the one I have drawn for you and try to think where you see that shape.

4 cm
1.8 cm
2 cm
2 cm

I see what you mean. I see that shape all around me. It's the shape of most roofs. It's also the shape of one of my favourite chocolate bars. I had no idea it was called a triangular prism. A fancy name for a common shape! How do I find the volume of that shape?

When I looked at the volume of a cuboid I said that it was the area of the rectangle × the depth of the solid.

So if the area of a triangle is ½ base × height, the volume of a triangular prism must be the area of the triangle × the length. Am I right?

Yes, well done. The volume of a triangular prism is simply the area of the triangular cross-section × the length. I'll use the formula to work out the volume of the triangular prism.

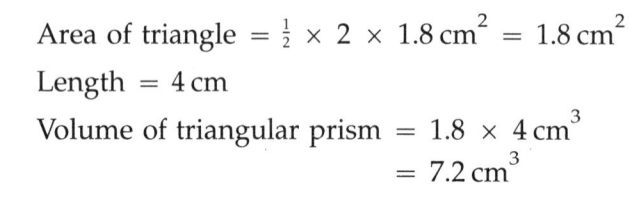

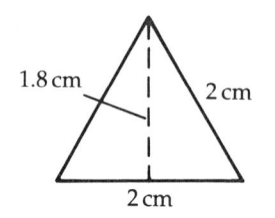

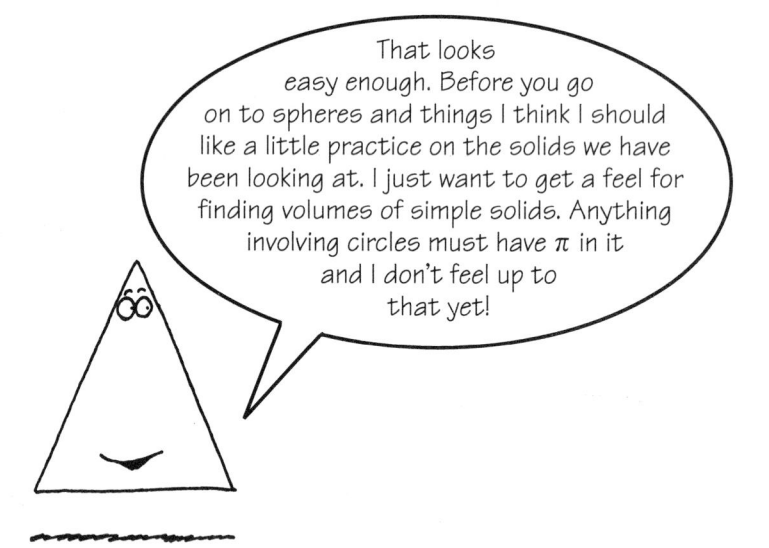

Area of triangle = $\frac{1}{2} \times 2 \times 1.8 \text{ cm}^2 = 1.8 \text{ cm}^2$

Length = 4 cm

Volume of triangular prism = $1.8 \times 4 \text{ cm}^3$
= 7.2 cm^3

1.8 cm, 2 cm, 2 cm

That looks easy enough. Before you go on to spheres and things I think I should like a little practice on the solids we have been looking at. I just want to get a feel for finding volumes of simple solids. Anything involving circles must have π in it and I don't feel up to that yet!

Try some for yourself

1. Complete the table.

	Length	Breadth	Depth	Volume
(a)	4 cm	4 cm	4 cm	
(b)	2.5 m	2.5 m	2.5 m	
(c)	21 mm	21 mm	21 mm	

2. Find the volume of these cuboids.

	Length	Breadth	Depth
(a)	15 mm	40 mm	3 mm
(b)	7.5 cm	7.5 cm	20 cm
(c)	23 cm	24 cm	50 cm

3. Find the volume of these triangular prisms.

	Base	Height	Length
(a)	4 cm	2 cm	10 cm
(b)	22 mm	10 mm	40 mm
(c)	6.8 m	4 m	22.4 m

I know we have to look at those solids involving circles. I am very nervous about formulae containing π. I expect I shall get all these volumes completely wrong! We have to make a start, which one are we looking at first?

I think the cylinder will be easier to look at first. It is shaped like a circle but has length like a cuboid. It's a very common shape so you really do need to know how to work out its volume.

I am not really sure what a cylinder looks like. I expect I know the shape, but not the name. Could you draw a cylinder for me, before we get involved with volumes? It helps to know what I am supposed to be thinking about.

That's quite reasonable. I am using mathematical language and assuming that you know what it means. When you see what a cylinder looks like you may be able to suggest how we can work out its volume.

Cylinders

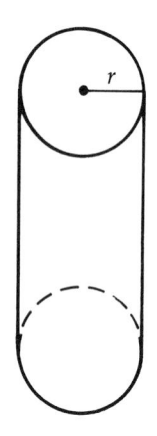

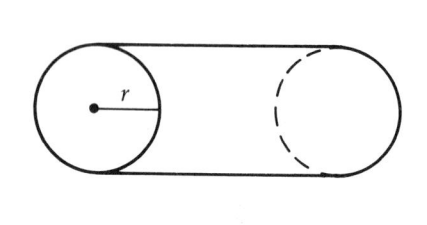

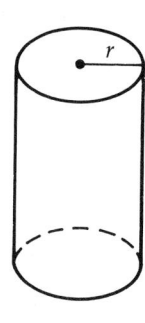

A cylinder is a very common shape, isn't it? All drink cans are cylinders, and cans of fruit, cat food, and . . . A lot of things are put into cylinders. I now know the correct mathematical word for a can. I'll amaze my friends.

That's all very well, but we are not just learning new words, we are supposed to be finding out about volumes. Do you remember how we found the volume of that triangular prism?

We found the area of the triangle at the end, then multiplied the area by the length. Do we do the same with the cylinder? The end is a circle so I'll need the formula for the area of a circle. Can you jog my memory, please?

You really do remember what we have already discussed. As you have suggested, to find the volume of a cylinder you multiply the area of the circular end by the length (or height) of the cylinder.

214

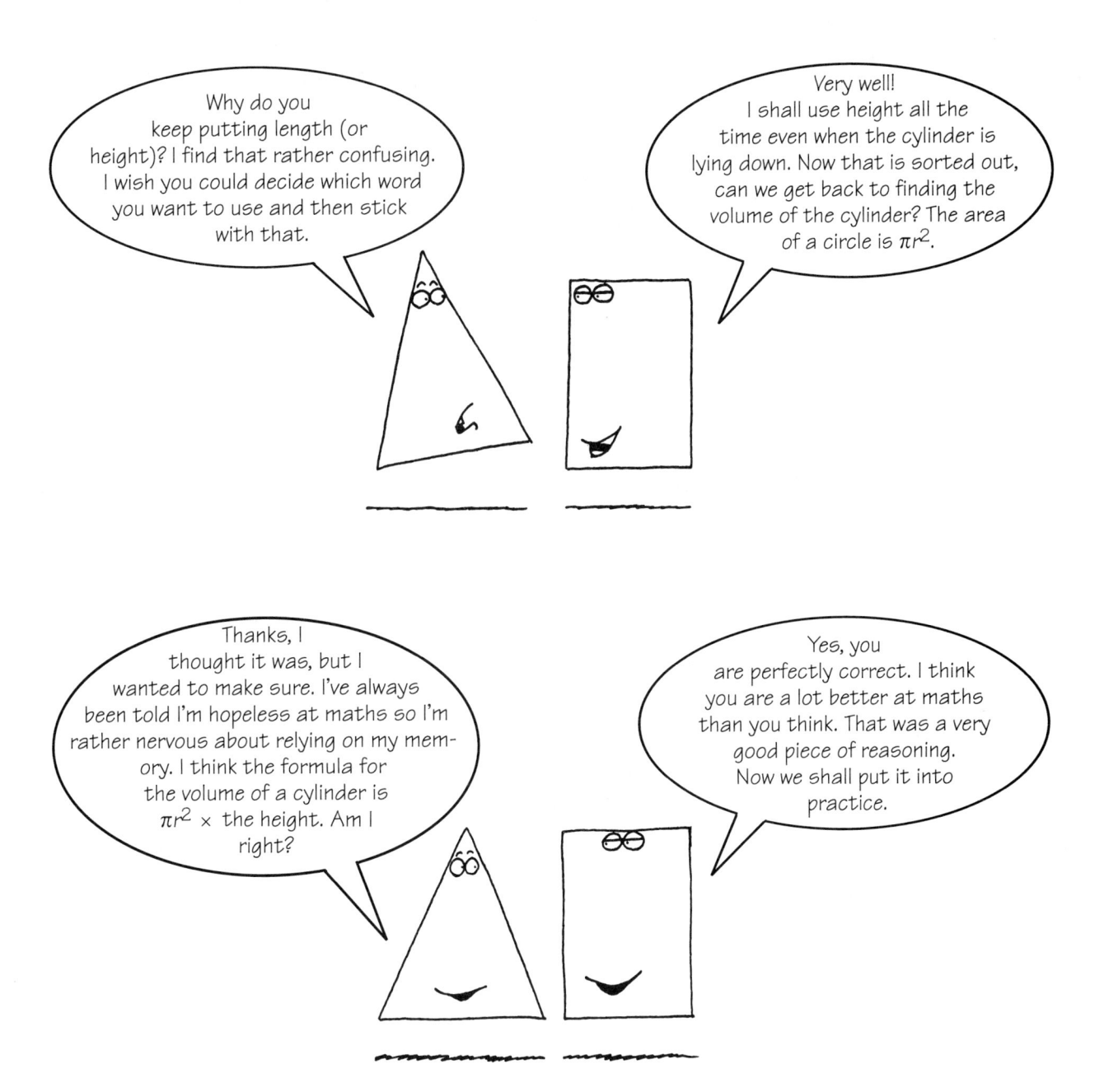

> **Why do you keep putting length (or height)? I find that rather confusing. I wish you could decide which word you want to use and then stick with that.**

> **Very well! I shall use height all the time even when the cylinder is lying down. Now that is sorted out, can we get back to finding the volume of the cylinder? The area of a circle is πr^2.**

> **Thanks, I thought it was, but I wanted to make sure. I've always been told I'm hopeless at maths so I'm rather nervous about relying on my memory. I think the formula for the volume of a cylinder is $\pi r^2 \times$ the height. Am I right?**

> **Yes, you are perfectly correct. I think you are a lot better at maths than you think. That was a very good piece of reasoning. Now we shall put it into practice.**

A can of baked beans has the following dimensions:

diameter, $d = 7.5$ cm
height, $h = 11$ cm

$$\text{Volume} = \pi r^2 h$$

We need to find the radius. The radius is half the diameter.

$$\begin{aligned}\text{Radius} &= 7.5 \div 2 = 3.75 \text{ cm} \\ \text{Volume} &= \pi \times 3.75^2 \times 11 \text{ cm}^3 \\ &= 485.965\,11 \text{ cm}^3 \\ &= 486 \text{ cm}^3 \text{ to 1 d.p.}\end{aligned}$$

Can you take it step by step? I know I have to work out the area of the circle but you went straight to the volume. Could you go through it again showing all the steps?

Diameter, $d = 7.5$ cm
Radius, $r = 3.75$ cm
Height, $h = 11$ cm

$$\begin{aligned}\text{Area of circle} &= \pi r^2 \\ &= \pi \times 3.75^2 \\ &= 44.178\,647 \text{ cm}^2 \\ &= 44.2 \times \text{cm}^2 \text{ to 1 d.p.}\end{aligned}$$

$$\begin{aligned}\text{Volume of cylinder} &= \text{Area of circle} \times \text{height} \\ &= 44.2 \times 11 \text{ cm}^3 \\ &= 486.2 \text{ cm}^3\end{aligned}$$

> Volume of cylinder = area of circular end × height

Let's find the volume of a pipe. The radius is 0.5 m. The height is 2 m.

$r = 0.5 \qquad h = 2$

$$\begin{aligned}\text{Area of circle} &= \pi \times 0.5 \times 0.5 \text{ m}^2 \\ &= 0.785 \text{ m}^2 \text{ to 3 d.p.}\end{aligned}$$

$$\begin{aligned}\text{Volume of pipe} &= \text{Area of circle} \times \text{height} \\ &= 0.785 \times 2 \text{ m}^3 \\ &= 1.57 \text{ m}^3 \text{ to 2 d.p.}\end{aligned}$$

Try some for yourself

4. All answers should be given to 2 decimal places.

   Find the volume of these cylinders.

   (a) Radius 5 cm, height 20 cm

   (b) Radius 1 m, height 4 m

   (c) Radius 8.5 cm, height 22 cm

5. Complete the table showing dimensions of cylinders.

Radius	Height	Volume
2	6	
3		339
	30	2356

> Have we finally finished with volumes? They seem rather pointless. All you learn is how much space is taken up by a solid. That's not much practical use.

> We haven't finished with volumes I'm afraid. We need to find out how we can use them. The most common use is to find out how much a container will hold. We work out its **capacity**, particularly for liquids.

Capacity

$$1 \text{ m}^3 = 1000 \text{ litres} = 1000\, l$$
$$1000 \text{ cm}^3 = 1 \text{ litre} = 1\, l$$
$$1 \text{ cm}^3 = 1 \text{ millilitre} = 1\, ml$$

A cylindrical oil drum has dimensions of 60 cm diameter, 90 cm height. How many litres of oil will it hold?

$d = 60$ cm radius, $r = 60 \div 2 = 30$ cm $h = 90$ cm

$$\begin{aligned}\text{Area of circle} &= \pi r^2 = 2827 \text{ cm}^2 \text{ to nearest whole number}\\ \text{Volume of cylinder} &= \pi r^2 h\\ &= 2827 \times 90 \text{ cm}^3\\ &= 254\,430 \text{ cm}^3\end{aligned}$$

$1000 \text{ cm}^3 = 1 \text{ litre}$

$254\,430 \div 1000 = 254.430$ litres
$ = 254$ litres to nearest litre

A rectangular tank is to be filled to the brim with water. How many litres of water will be needed to fill the tank?

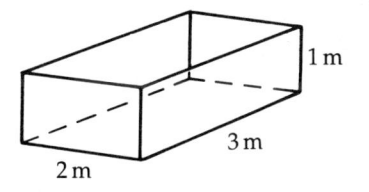

$$\begin{aligned}\text{Volume of a cuboid} &= l \times w \times h\\ &= 3 \times 2 \times 1 \text{ m}^3\\ &= 6 \text{ m}^3\end{aligned}$$

$$1 \text{ m}^3 = 1000 \text{ litres}$$
$$6 \text{ m}^3 = 6000 \text{ litres}$$

6000 litres of water will be needed to fill the tank.

Nets

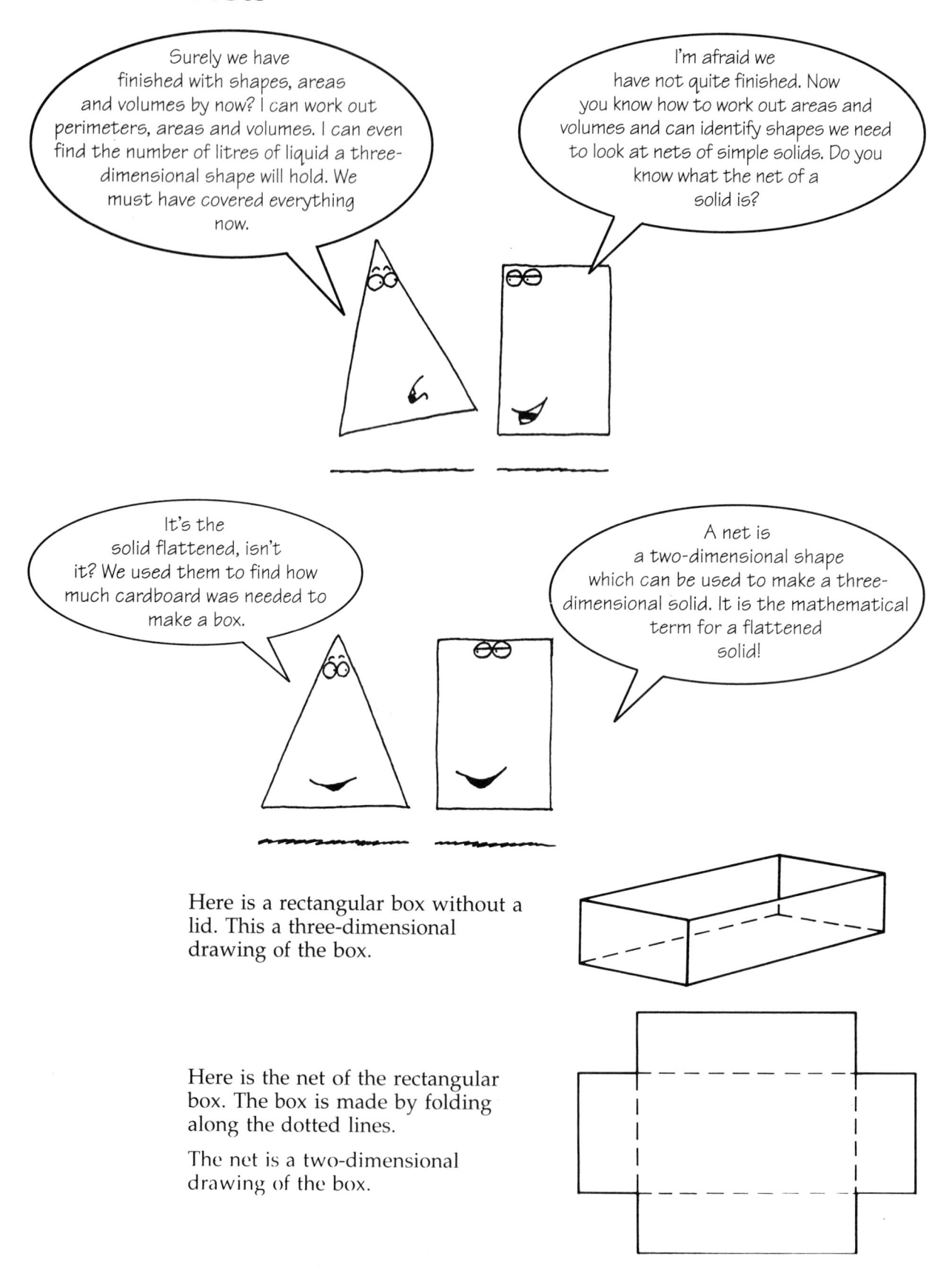

Surely we have finished with shapes, areas and volumes by now? I can work out perimeters, areas and volumes. I can even find the number of litres of liquid a three-dimensional shape will hold. We must have covered everything now.

I'm afraid we have not quite finished. Now you know how to work out areas and volumes and can identify shapes we need to look at nets of simple solids. Do you know what the net of a solid is?

It's the solid flattened, isn't it? We used them to find how much cardboard was needed to make a box.

A net is a two-dimensional shape which can be used to make a three-dimensional solid. It is the mathematical term for a flattened solid!

Here is a rectangular box without a lid. This a three-dimensional drawing of the box.

Here is the net of the rectangular box. The box is made by folding along the dotted lines.

The net is a two-dimensional drawing of the box.

This is a cube of dimensions 2 cm × 2 cm × 2 cm. The cube has six faces, each a square measuring 2 cm × 2 cm.

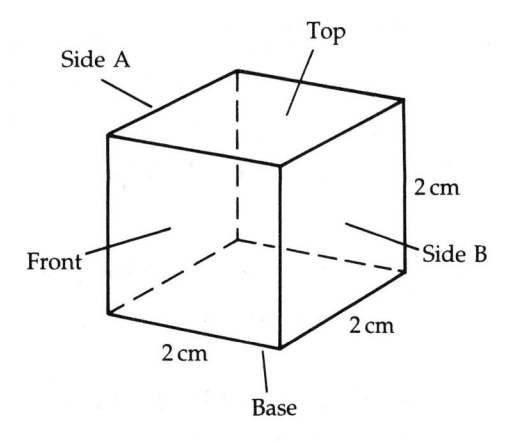

This is a net of the cube. Each square face measures 2 cm by 2 cm.

The total area of the cube is
$6 \times (2 \times 2) \text{ cm}^2$
$= 6 \times 4 \text{ cm}^2$
$= 24 \text{ cm}^2$

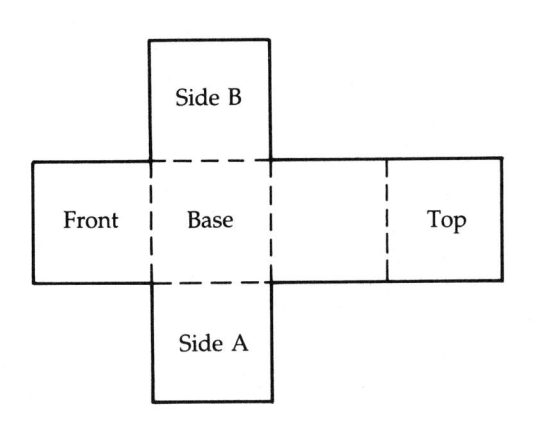

A cardboard cake box measuring 12 cm × 10 cm × 2 cm is to be made from a sheet of card. The card measures 20 cm × 30 cm.

(a) Find the area of the net of the cake box.
(b) Find the amount of card remaining when the box has been cut out.

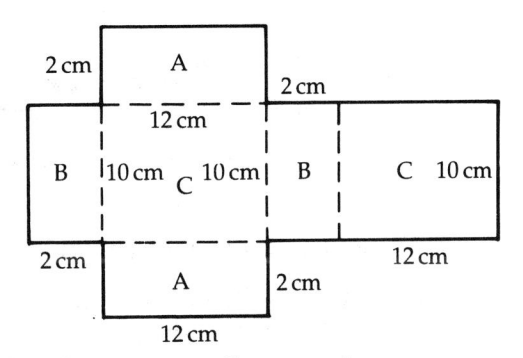

(a) To find the area of the net we must break it into small rectangles.

There are 2 rectangles measuring 12 cm × 2 cm. Label these A.
There are 2 rectangles measuring 10 cm × 2 cm. Label these B.
There are 2 rectangles measuring 12 cm × 10 cm. Label these C.

Find the area of rectangle A.
$2 \times 12 = 24 \text{ cm}^2$

Find the area of rectangle B.
$2 \times 10 = 20 \text{ cm}^2$

Find the area of rectangle C.
$12 \times 10 = 120 \text{ cm}^2$

219

Find the area of the cake box.

Area = 2 × rectangle A + 2 × rectangle B + 2 × rectangle C

Area = (2 × 24 + 2 × 20 + 2 × 120) cm^2

= (48 + 40 + 240) cm^2

= 328 cm^2

(b) To find the amount of card remaining.

Area of sheet of card = 20 × 30 cm^2
= 600 cm^2

Area of card remaining when the net of the box has been cut out is
(600 − 320) cm^2 = 280 cm^2

> I can see that knowing about the nets of cubes and cuboids will be useful. Lots of boxes are made in those shapes. I hope I do not have to know about the nets of the other solids. They will be much more difficult.

> Certainly, the nets of cubes and cuboids are the most useful ones to know. I think it will be a good idea to look at one more net – the net of the triangular prism. We know that roofs and chocolate bars take that shape.

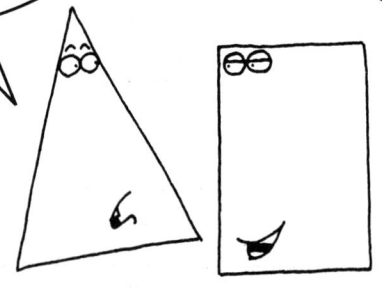

Here is the net of a triangular prism.

It is made up of a large rectangle and two identical triangles.

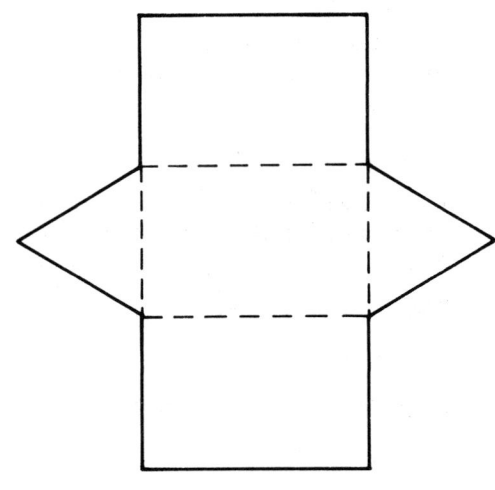

Area of the net of a triangular prism = Area of rectangle + 2 × area of triangular end

The cardboard model of a house has its roof section made as a triangular prism.

The rectangle is 15 cm × (5 + 5 + 5) cm.

The triangle has base 5 cm and height 4.3 cm.

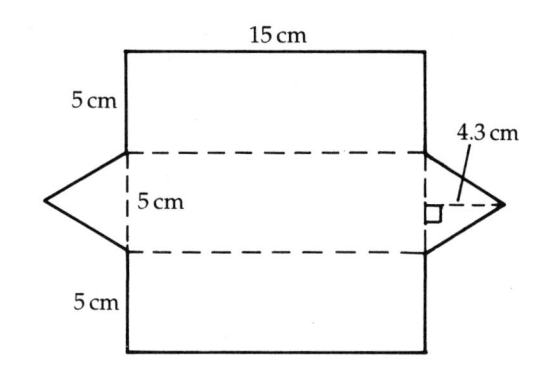

Area of rectangle = 15 × 15 cm² = 22.5 cm²   (length × width)

Area of triangle = $\frac{1}{2}$ × 5 × 4.3 = 10.75 cm²   ($\frac{1}{2}$ base × height)

Area of 2 triangles = 2 × 10.75 cm² = 21.50 cms

Area of net = area of rectangle + 2 × area of triangular end
= 225 cm² + 21.50 cm²
= 246.50 cm²

▽ Try some for yourself

6. Draw the nets of the following solids.
   (a) A cube 2 × 2 × 2
   (b) A cuboid 10 × 2 × 12
   (c) A triangular prism base 6 cm, height 5.2 cm, length 10 cm.

7. Which of the diagrams are nets of a cube?
   (a)   (b)   (c)

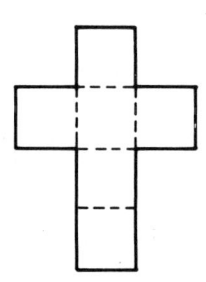

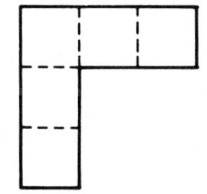

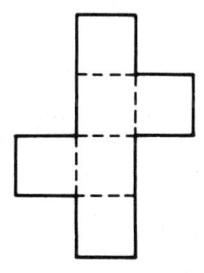

8. A cake box is made from thin card. The dimensions are 20 cm × 20 cm × 7.5 cm.
   (a) Draw the net of the box which includes the lid.
   (b) Calculate the area of cardboard needed to make the box.

9. The numbers on the opposite faces of a die add up to 7.

(a) 6 + 1 = 7

Complete the faces of the die making sure that opposite faces total 7.

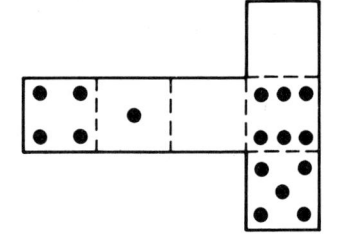

(b) Complete the faces of the dice. Make sure that opposite faces total 7.

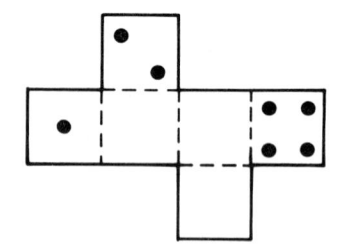

What have you learnt about volumes and nets?

I was frightened to be asked to work out volumes. Now I know that I can work them out as long as I have the formulae. Nets are easy-well, some are!

- ✓ I know that volume is the measure of the space filled by a solid object.
- ✓ I know that volume is always measured in cubic units.
- ✓ I know the names of common solids.
- ✓ I can find the volumes of cubes, cuboids, triangular prisms and cylinders when I know the formulae.
- ✓ I know how to use volumes to calculate capacity.
- ✓ I know what the net of a solid is.
- ✓ I can draw the nets of cubes, cuboids and triangular prisms.
- ✓ I can find the areas of nets.

Further Questions

1. Find the volume of air in each of the following.

		Width	Length	Height
(a)	Sports hall	18 m	30 m	8 m
(b)	Telephone kiosk	0.9 m	0.9 m	2.4 m
(c)	Shower cubicle	1.8 m	0.8 m	2.8 m

2. A swimming pool is rectangular in shape. The length is 24 m and the width 19 m. The depth is a constant 2 m.

   (a) Calculate the volume of the pool.

   (b) Calculate the amount of water needed to fill the pool to the brim.

3. A straight-sided glass has diameter 6 cm and height 10 cm. The glass is half full of water. Calculate the volume of water in the glass. How many litres of water are in the glass?

4. A soft drink can has diameter 6.4 cm and height 12 cm.

   (a) Find the volume of the can.

   (b) How much drink could the can hold?

   (c) The can contains 325 ml. Is it completely full?

5. A large sports hall has the following dimensions.

   Length = 72 m
   Width = 48 m
   Volume = 53 000 m^3

   Find the height of the hall.

6. A cylindrical water dispenser has a radius of 13 cm and a height of 30 cm.

   (a) Calculate the volume of the dispenser.

   (b) Find the number of litres of water required to fill the dispenser.

7. An iron roller is cylindrical in shape. Its diameter is 50 cm. The height is 125 cm.

   Find the volume of the roller.

8. A chest has a volume of 7005 cm^3. The height is 9.7 cm and the width 33 cm. Find the length of the chest.

9. A child's ridge tent is in the shape of a triangular prism.

   Height = 90 cm
   Width = 120 cm
   Length = 190 cm

   Find the volume of space inside the tent.

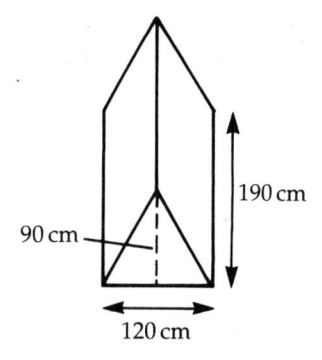

10. A two-man ridge tent is in the shape of a triangular prism.

    The dimensions of the triangle are shown in the diagram.

    The tent is 190 cm long.

    Find the volume of air inside the tent.

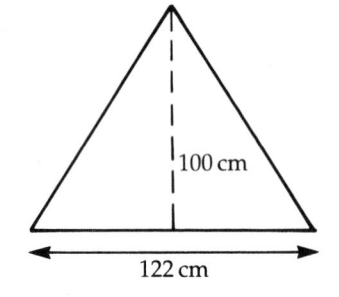

11. A tea box is in the shape of a cuboid. The dimensions are 7.5 cm × 7.5 cm × 9.5 cm.

    (a) Draw the net of the box, including the lid.
    (b) Find the area of card needed to make the box.

12. A chocolate bar is made in the shape of a triangular prism. The bar comes in three sizes.

    Fun size   : base = 1 cm, height = 0.9 cm, length = 8 cm
    Standard size : base = 2 cm, height = 1.7 cm, length = 16 cm
    Giant size   : base = 4 cm, height = 3.5 cm, length = 32 cm

    (a) Draw a net for each bar indicating measurements.
    (b) Find the amount of silver paper needed to wrap one bar of each size.

Solutions

1. (a) $4 \times 4 \times 4 = 64 \, cm^3$
   (b) $2.5 \times 2.5 \times 2.5 = 15.625 \, cm^3$
   (c) $21 \times 21 \times 21 = 9261 \, cm^3$

2. (a) $15 \times 40 \times 3 \, mm^3 = 1800 \, mm^3$
   (b) $7.5 \times 7.5 \times 20 \, cm^3 = 1125 \, cm^3$
   (c) $23 \times 24 \times 50 \, cm^3 = 27\,600 \, cm^3$

3. (a) $\frac{1}{2} \times 4 \times 2 \times 10 \, cm^3 = 40 \, cm^3$
   (b) $\frac{1}{2} \times 22 \times 10 \times 40 \, mm^3 = 4400 \, mm^3$
   (c) $\frac{1}{2} \times 6.8 \times 4 \times 22.4 \, m^3 = 304.6 \, m^3$

4. (a) $\pi r^2 h = \pi \times 5^2 \times 20 \, cm^3 = 1570.80 \, cm^3$ to 2 d.p.
   (b) $\pi r^2 h = \pi \times 1^2 \times 4 \, m^3 = 21.57 \, m^3$ to 2 d.p.
   (c) $\pi r^2 h = \pi \times 8.5^2 \times 22 \, cm^3 = 4993.56 \, cm^3$ to 2 d.p.

5.

Radius	Height	Volume
2	6	**75.4**
3	**12**	339
5	30	2356

6. (a) [net of cube with side 2]

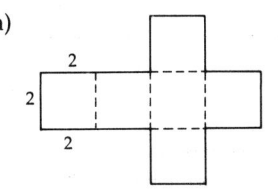

(b) [net of rectangular box 10 × 2 × 12]

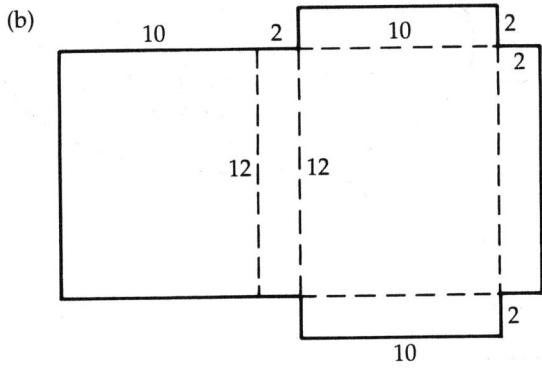

(c) [net with 6 cm and 5.2 cm dimensions]

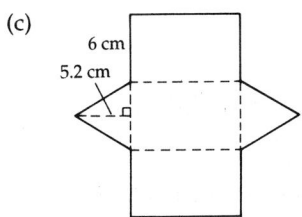

7. (a) Yes (b) No (c) Yes

8. (a) [net of box: 20 cm × 20 cm faces with 7.5 cm sides]

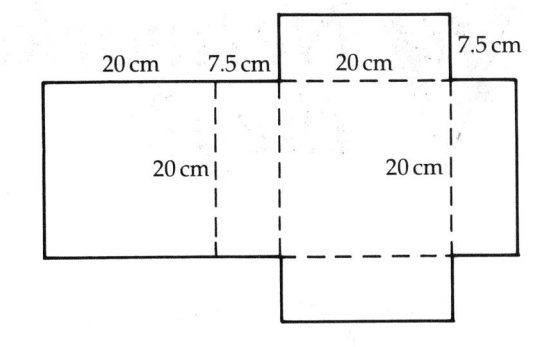

(b) $2 \times 20 \times 20 = 800$
$4 \times 20 \times 7.5 = 600$
Total area of cardboard $= \overline{1400}$ cm^2

9. (a) [die net with dots]

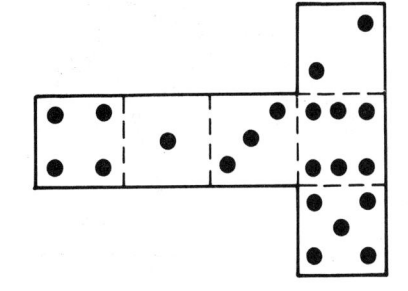

(b) [die net with dots]

225

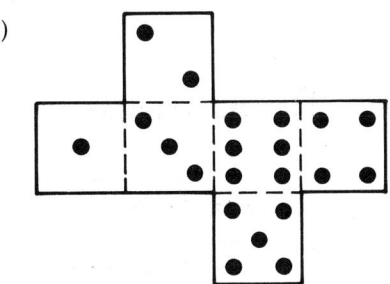

14 Networks – what are they?

What is all this about networks? I never learnt about them at school.

A network is any diagram where lines indicate links between points or items. It's a blanket word to cover all sorts of relationships. I'll show you some examples of networks.

Road maps

A road map is a network. It uses lines to represent the roads linking different places.

This road map is drawn to scale so that distances and directions can be found.

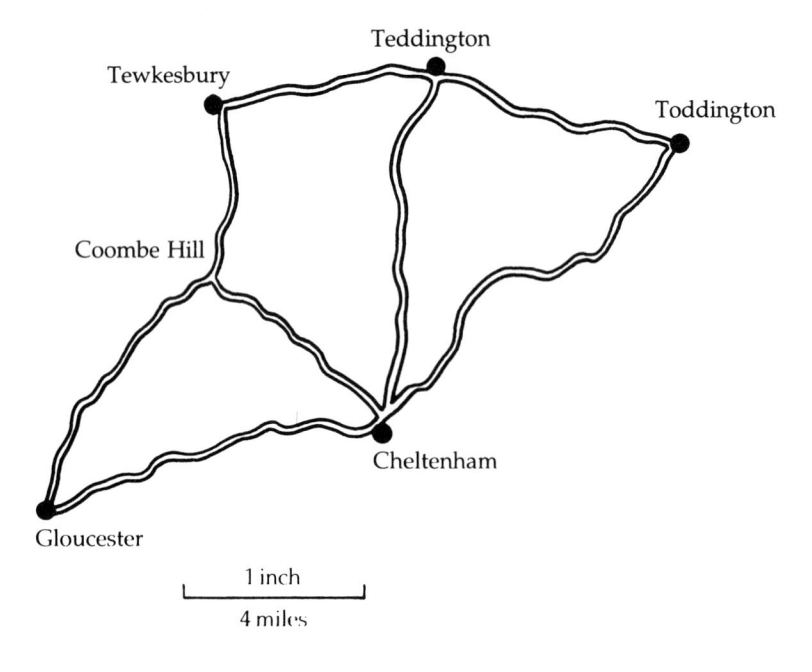

1 inch
4 miles

That's just an ordinary road map. How do I change it into one of these networks?

*It's very easy to redraw the map as a network. You need to mark the distance between places and give a rough idea of direction — then you will have a **network**.*

Here is a possible network. The distances are marked and we are given a rough idea of direction.

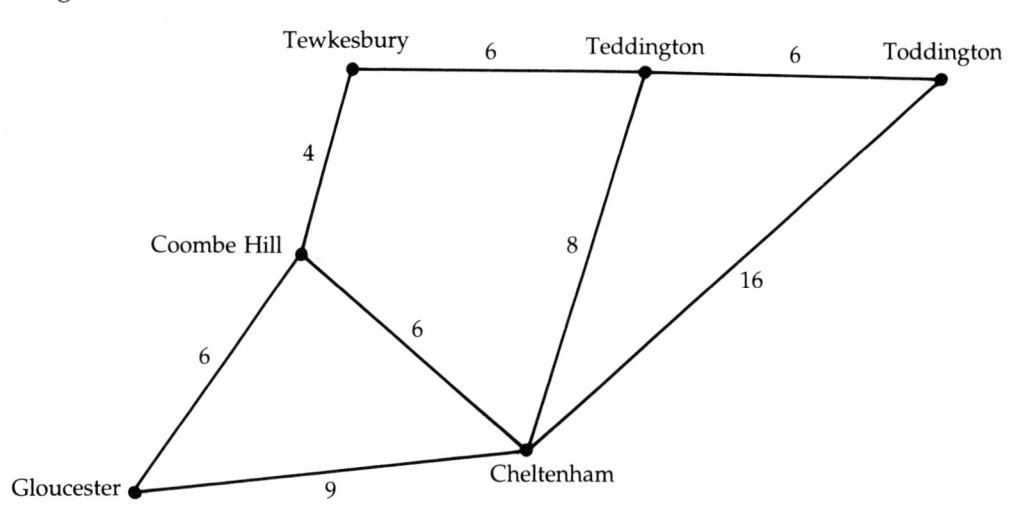

- Tewkesbury — 6 — Teddington — 6 — Toddington
- Tewkesbury — 4 — Coombe Hill
- Coombe Hill — 6 — Cheltenham
- Teddington — 8 — Cheltenham
- Toddington — 16 — Cheltenham
- Gloucester — 6 — Coombe Hill
- Gloucester — 9 — Cheltenham

O.K. I can see how to produce a network diagram from a road map. What now? Why am I doing it?

That's a good question. You need to know why we do this or it's meaningless. One of the ways we can use networks is to help us to work out distances for different routes through the network.

We are often asked to find the shortest (or longest) distance between two places.

Let's look at the different routes we can take to get from Cheltenham to Gloucester. We can also work out the different distances involved.

Cheltenham–Gloucester	9 miles
Cheltenham–Coombe Hill–Gloucester	12 miles
Cheltenham–Teddington–Tewkesbury–Gloucester	24 miles
Chteltenham—Toddington–Teddington–Tewkesbury–Gloucester	38 miles

We can see that the shortest route is 9 miles and that involves going direct from Cheltenham to Gloucester.

We can see that the longest route is 38 miles. This takes us through all the places on the map – the scenic route.

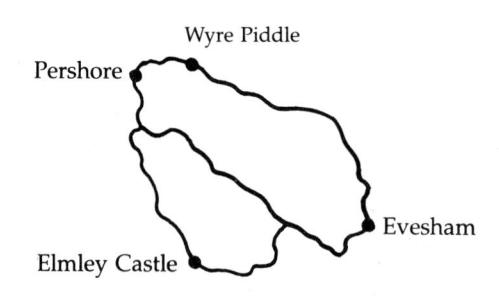

That looks quite easy. I think I might manage to do a question on my own. Is there one I can try?

Yes. I have a map of the roads between Pershore and Evesham. The scale is 4 miles to 1 inch.

▽ Try some for yourself
1.

(a) Draw a simplified network showing the roads. Mark the estimated distances between places.

(b) Use your network to find the shortest route from Evesham to Pershore.

(c) Use your network to find the longest route.

Transport networks

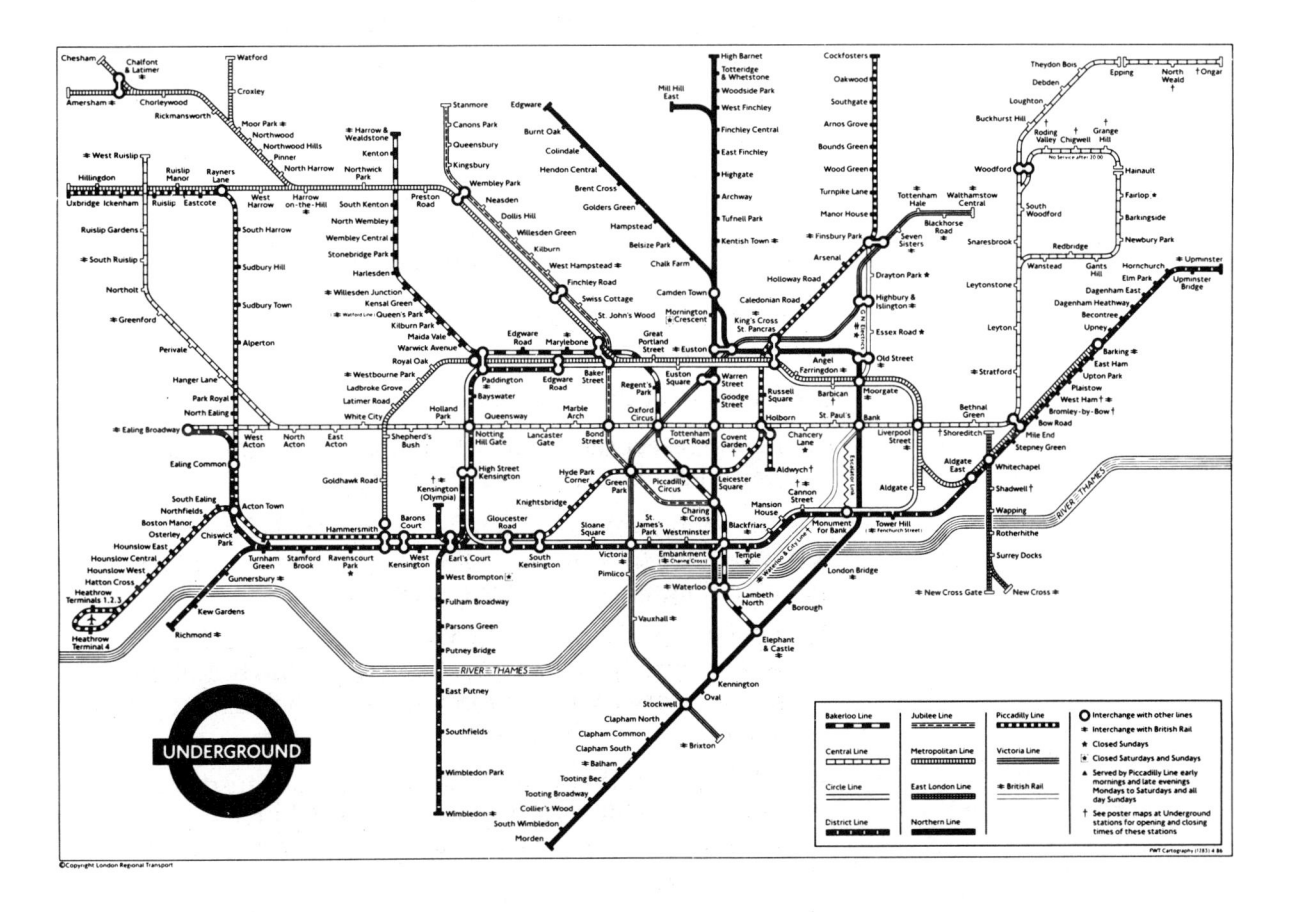

> Road networks are just one example of transport networks. There are rail networks, canal networks, air networks and bus networks. Here are some familiar networks.

This familiar map of the London Underground is a network diagram.

229

This diagram showing the routes of local buses travelling between Cheltenham and Gloucester is a network diagram.

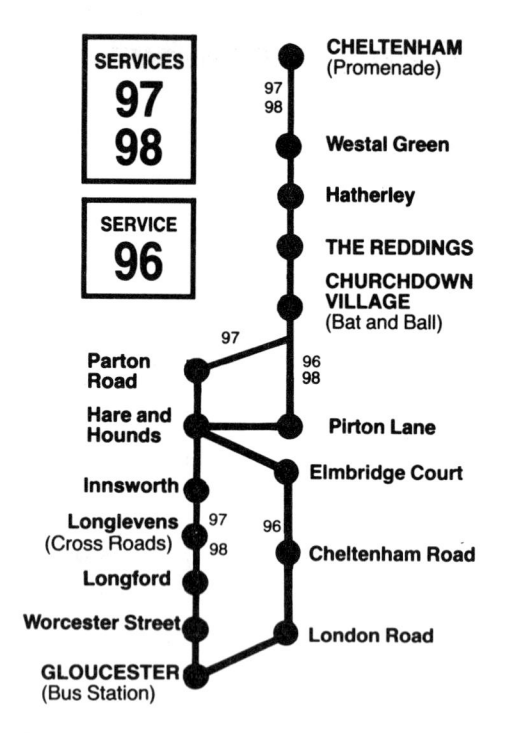

SERVICES 97 98

SERVICE 96

- CHELTENHAM (Promenade)
- Westal Green
- Hatherley
- THE REDDINGS
- CHURCHDOWN VILLAGE (Bat and Ball)
- Parton Road
- Hare and Hounds
- Pirton Lane
- Elmbridge Court
- Innsworth
- Longlevens (Cross Roads)
- Cheltenham Road
- Longford
- Worcester Street
- London Road
- GLOUCESTER (Bus Station)

Neither of these show distances but they indicate routes and stations/stops along those routes.

▽ Try some for yourself

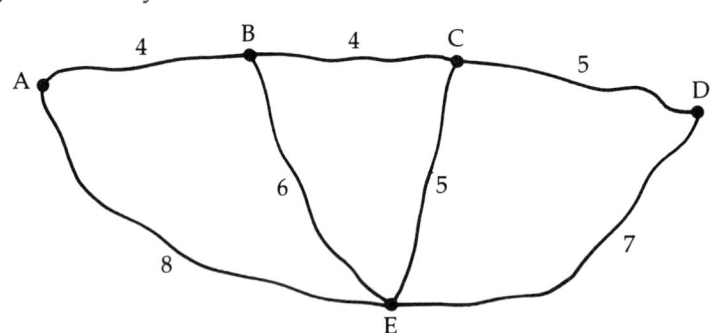

2.

In this road system the distances between points are marked in kilometres. Find:

(a) the shortest distance from A to D.

(b) the longest route from A to D that does not require any road to be covered more than once.

3. The following are route details for the 94, 94B and 95 but services between Cheltenham and Gloucester. Produce a network diagram to illustrate these services.

94B Cheltenham Promenade–Westal Green–Lansdown–Arle Court– Bamfurlong– Staverton–Dowty Rotol–Churchdown–Greyhound Gardens–Cheltenham Road–London Road–Gloucester

94 Cheltenham Promenade–Westal Green–Lansdown–Arle Court–Staverton–Dowty Rotol–Churchdown–Greyhound Gardens–Cheltenham Road–London Road–Gloucester

95 Cheltenham Promenade–Westal Green–Lansdown–Arle Court–Golden Valley By-Pass–Greyhound Gardens– Cheltenham Road–London Road–Gloucester

Relationship networks

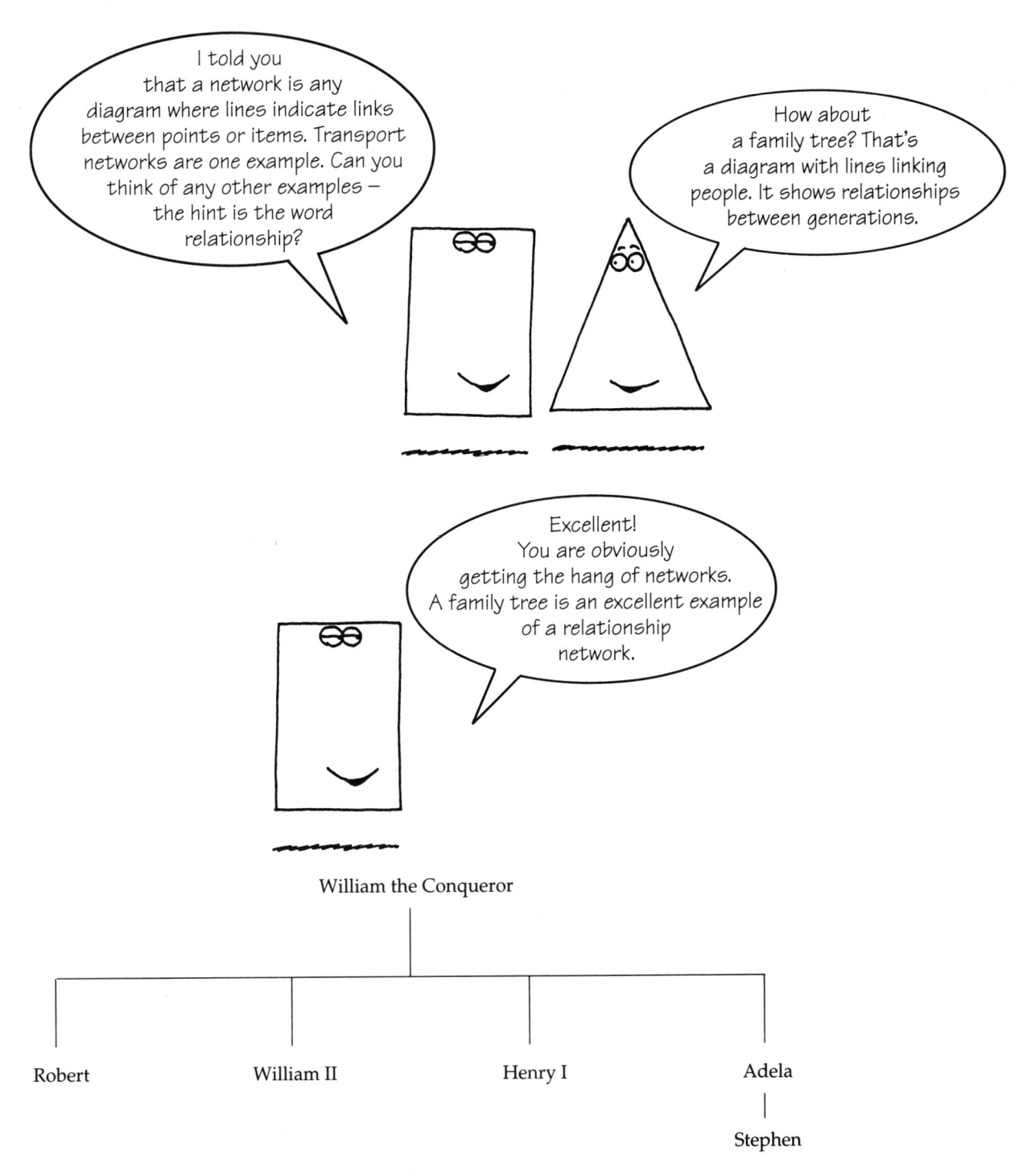

I told you that a network is any diagram where lines indicate links between points or items. Transport networks are one example. Can you think of any other examples – the hint is the word relationship?

How about a family tree? That's a diagram with lines linking people. It shows relationships between generations.

Excellent! You are obviously getting the hang of networks. A family tree is an excellent example of a relationship network.

```
                    William the Conqueror
          ┌──────────────┼──────────────┬──────────┐
        Robert       William II       Henry I     Adela
                                                    │
                                                 Stephen
```

This is the family tree of William the Conqueror. Here the lines do not show direction but descent.

The lines mean 'is parent of'.

Thus William is parent of Robert, William II, Henry I and Adela. Adela is parent of Stephen.

231

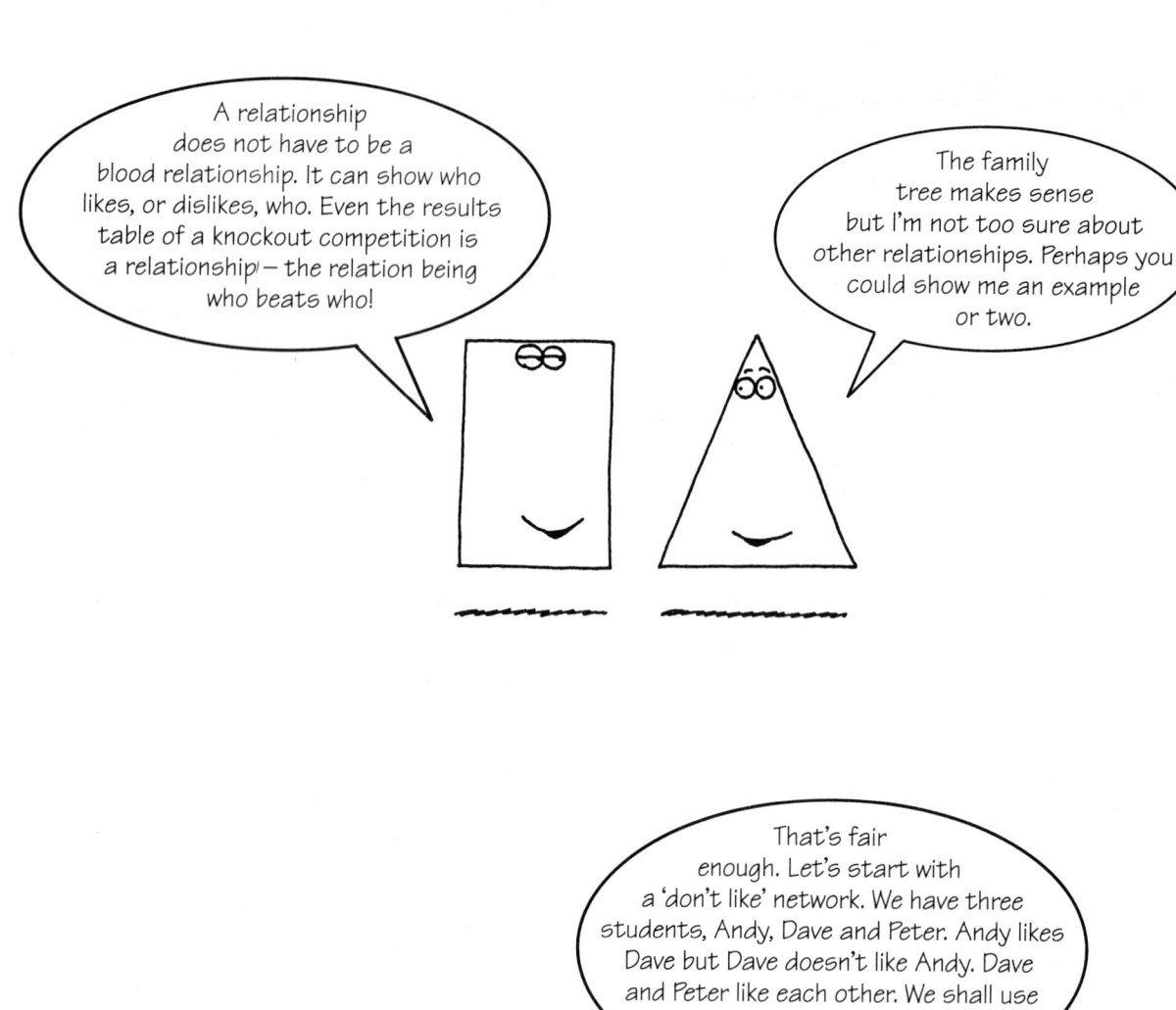

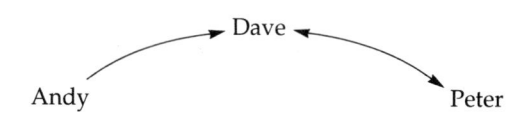

> A relationship does not have to be a blood relationship. It can show who likes, or dislikes, who. Even the results table of a knockout competition is a relationship — the relation being who beats who!

> The family tree makes sense but I'm not too sure about other relationships. Perhaps you could show me an example or two.

> That's fair enough. Let's start with a 'don't like' network. We have three students, Andy, Dave and Peter. Andy likes Dave but Dave doesn't like Andy. Dave and Peter like each other. We shall use an arrow → to indicate 'likes'.

Andy → Dave ← Peter (with arrow from Dave to Peter)

What does this tell us about poor Andy?

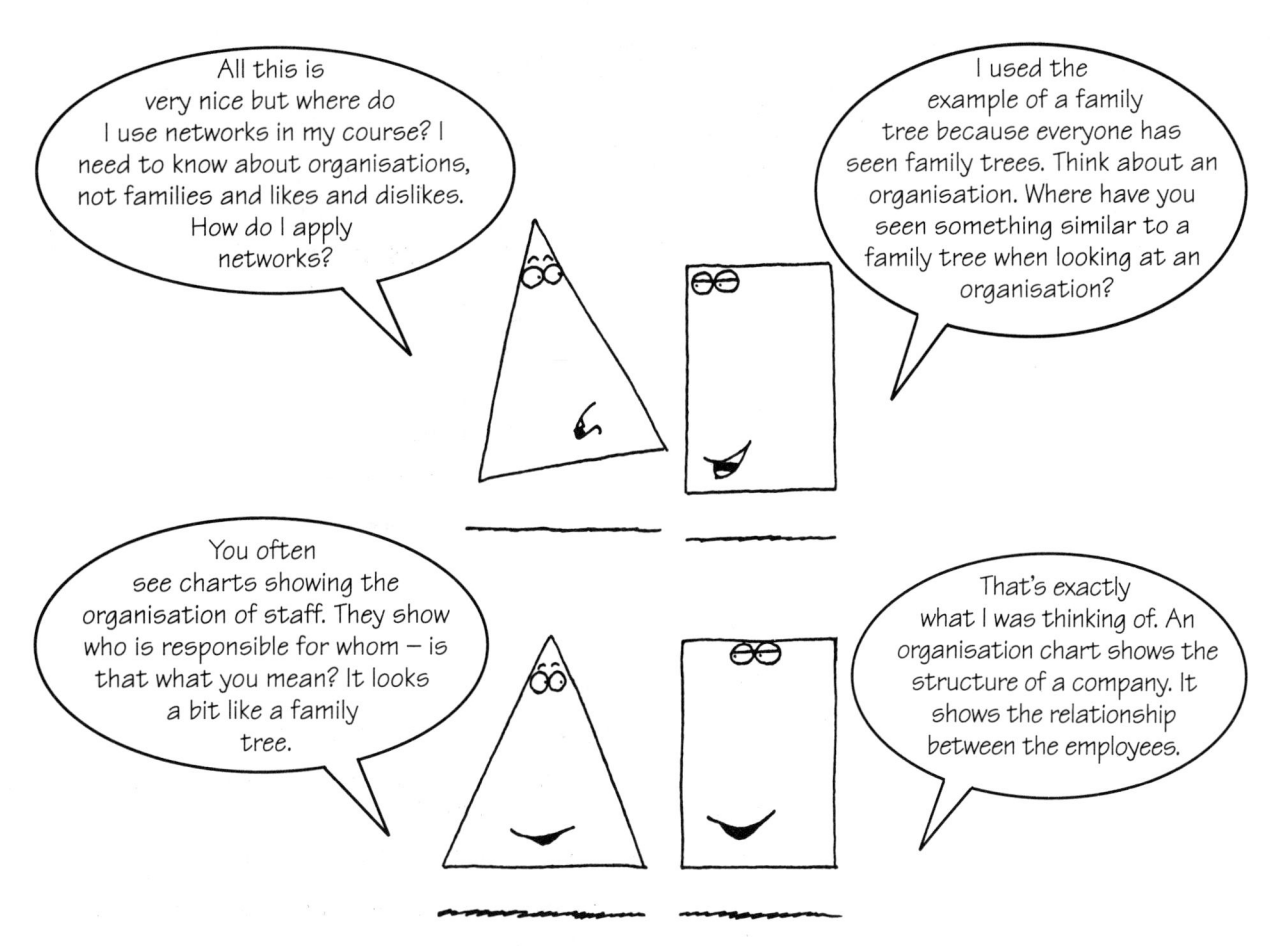

All this is very nice but where do I use networks in my course? I need to know about organisations, not families and likes and dislikes. How do I apply networks?

I used the example of a family tree because everyone has seen family trees. Think about an organisation. Where have you seen something similar to a family tree when looking at an organisation?

You often see charts showing the organisation of staff. They show who is responsible for whom – is that what you mean? It looks a bit like a family tree.

That's exactly what I was thinking of. An organisation chart shows the structure of a company. It shows the relationship between the employees.

A small shop employs three sales assistants who work under the management of the assistant manager. The assistant manager is responsible to the manager who, in turn, is responsible to the owner.

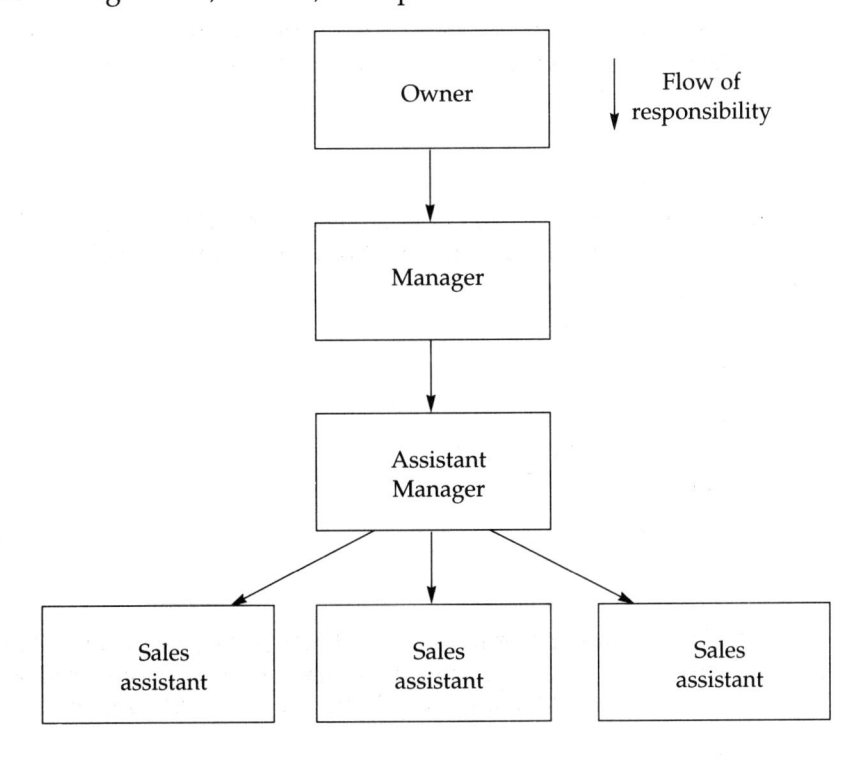

```
           Owner              ↓ Flow of
             │                  responsibility
             ▼
          Manager
             │
             ▼
         Assistant
          Manager
         ╱   │   ╲
        ▼    ▼    ▼
     Sales  Sales  Sales
   assistant assistant assistant
```

> I have given you some examples of relationship networks. Now it's time to try to do some on your own. That's the best way to learn, you know.

Try some for yourself

4. (a) Draw a network to show this family tree.

Ben has three sons, Martin, John and Fergus. Only Fergus has a child, a son Colin.

The network should indicate 'is father of'.

(b) Draw a second network for the same famly where the arrow means 'is brother of'.

5. Draw a network to show the relationship between Angela, Ben and Caroline. Angela likes Ben and Caroline. Ben likes Caroline, Caroline likes Angela.

6. Use the arrow to mean beats when you draw a network to represent the results in this knockout tennis match.

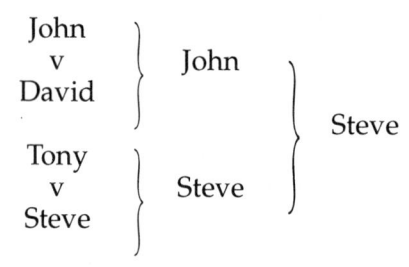

```
John  ⎫
  v   ⎬ John  ⎫
David ⎭        ⎬ Steve
Tony  ⎫        ⎪
  v   ⎬ Steve ⎭
Steve ⎭
```

7. On successive Saturdays four football teams played each other as shown on the next page. The results are indicated by the network diagram where the arrow means 'beats'.

	Home	**Away**
October 1st	Bourton Prestbury	Winchcombe Leckhampton
October 8th	Winchcombe Leckhampton	Prestbury Bourton

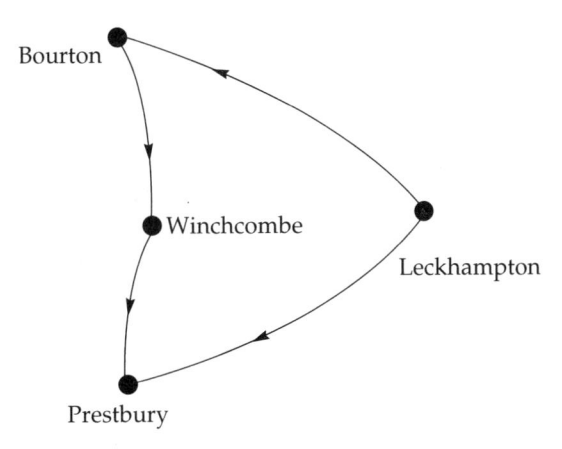

If an away win was worth 3 points and a home win 2 points, what were each team's total points?

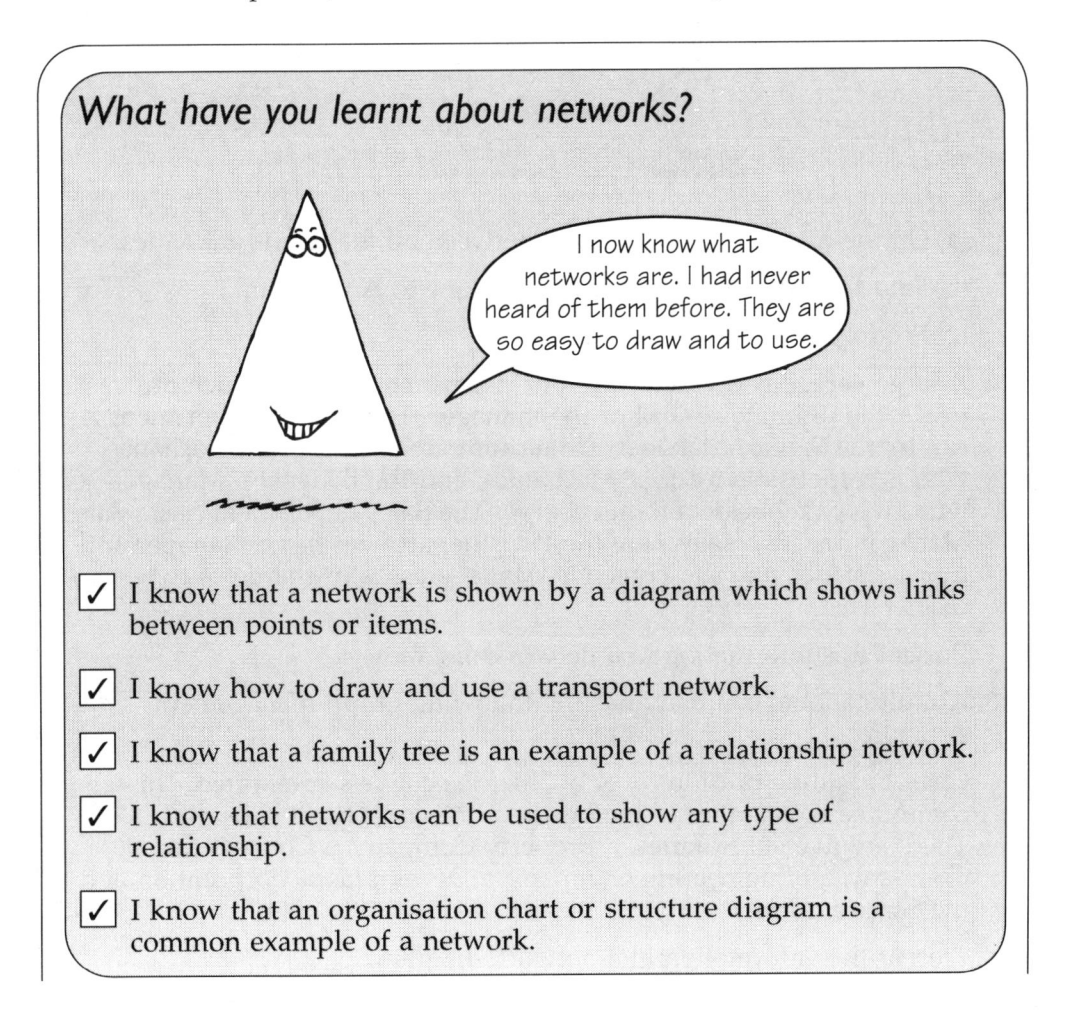

What have you learnt about networks?

I now know what networks are. I had never heard of them before. They are so easy to draw and to use.

- [✓] I know that a network is shown by a diagram which shows links between points or items.
- [✓] I know how to draw and use a transport network.
- [✓] I know that a family tree is an example of a relationship network.
- [✓] I know that networks can be used to show any type of relationship.
- [✓] I know that an organisation chart or structure diagram is a common example of a network.

Further Questions

1.

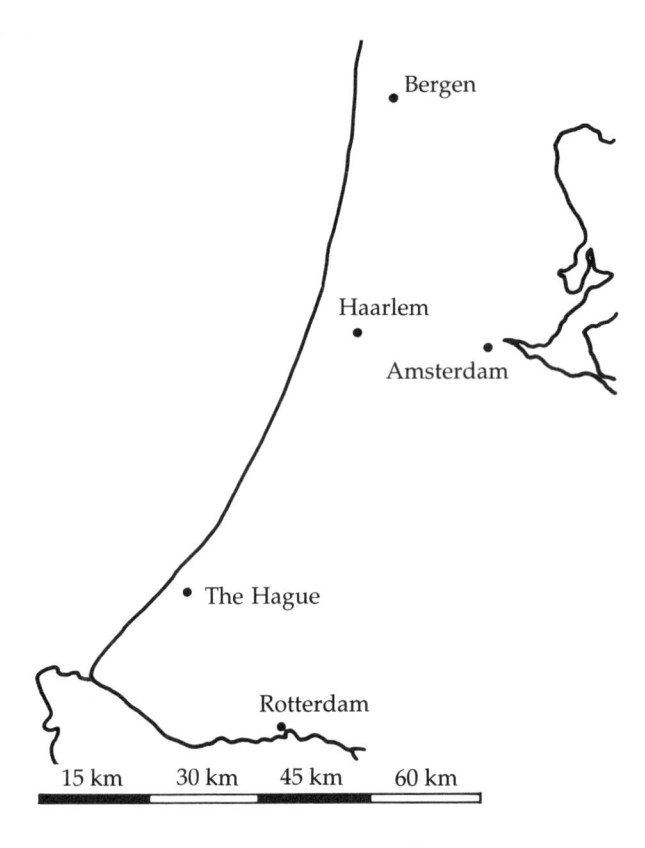

(a) On the map indicate distances between all the towns and cities.

(b) Find the longest route from The Hague to Amsterdam.

(c) Produce a simplified network.

2. A large garage is divided into four departments, all of which are under the ultimate control of the manager. The Service department is run by the Service Manager. Under him are the Foreman mechanic, who is responsible for four mechanics, and the Bodyshop Manager with two panel beaters in his charge. The Sales department has a Sales Manager and two salesmen; the Parts department has a Manager and two assistants, and the Forecourt Manageress is in charge of two cashiers.

Present this information in a network diagram.

3. A multinational company has the following group management structure:

Chief Executive presides over a Chief Executive's committee. This committee is made up of the heads of the constituent subsidiaries. There are five subsidiaries: a Property Company, a Construction Company, an Engineering Company, a Development Corporation and the Services Group.

Present this information in a network diagram.

4. GLOUCESTERSHIRE AIRPORT
Schedule Timetable

DEP	ARR	DAYS	DEP	ARR
TO DUBLIN			**FROM DUBLIN**	
0810	0925	MTWTF	0955	1150
1545	1740	MTWTF	1810	1925
TO JERSEY			**FROM JERSEY**	
0930	1050	SS	1110	1210
1100	1200	MTF	1215	1335
1230	1330	SS	1445	1605
1755	1855	MTWTF	0800	0900
TO GUERNSEY			**FROM GUERNSEY**	
0930	1025	SS	1510	1605
1100	1230	MTF	1240	1335
1755	1915	MTWTF	0740	0900
TO WATERFORD			**FROM WATERFORD**	
1200	1310	MWF	1425	1535
TO GLASGOW			**FROM GLASGOW**	
0745	0945	MTWTF	1720	1920

(a) Use the information in the timetable to produce the following simplified networks. In each case, indicate the flying time.

  (i) Weekday morning flights departing Staverton up to 12 noon.

  (ii) Weekday morning flights arriving Staverton up to 12 noon.

  (iii) Weekday afternoon flights departing Staverton.

  (iv) Weekday afternoon flights arriving Staverton after 12 noon.

(b) From the flying times explain your deductions concerning routes for the airlines.

Map of Great Britain and Northern France

5. The College of Midshire is an FE college running a number of GNVQ courses. The college structure is such that each GNVQ course comes within its own school. There are the schools of:
   Leisure and Tourism, and Business Studies,
   Hospitality and Catering, and Health and Social Care,
   Built Environment, and Manufacturing,
   Science and Information Technology.

These eight schools are organised into four faculties:
   Faculty of Business and Tourism,
   Faculty of Food and Care,
   Faculty of Engineering,
   Faculty of Science and Computing.

The four Heads of faculty report to the appropriate Vice-Principal:
   Vice-Principal with responsibility for Business and Service,
   Vice-Principal with responsibility for Technology.

The two Vice-Principals are directly responsible to the Principal.

Draw an organisation chart to display this information.

If abreviations are used a key must be provided.

Solutions

1. (a) [graph: Pershore–Wyre Piddle 2, Pershore–Elmley Castle 4, Pershore–Evesham 5, Wyre Piddle–Evesham 6, Elmley Castle–Evesham 4, Evesham–Pershore 2]

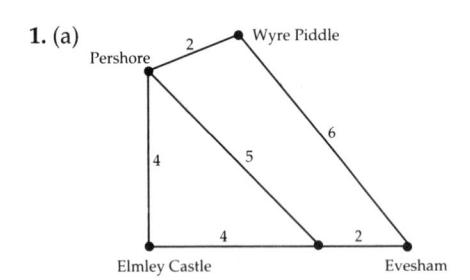

   (b) 7 miles

   (c) 10 miles

2. (a) A → B → C → D; 13 kms

   (b) A → E → B → C → E → D; 30 kms

3. [route diagram: Cheltenham — 94 — Westal Green — 94B — Lansdown — 95 — Arle Court — (94, 94B via Staverton, Dowty Rotal, Churchdown, Bansfurlong; 95 Direct via Golden Valley By-pass) — 94, 94B — Greyhound Gardens — 14.16 — Cheltenham Road — 94, 94B, 95 — London Road — Gloucester]

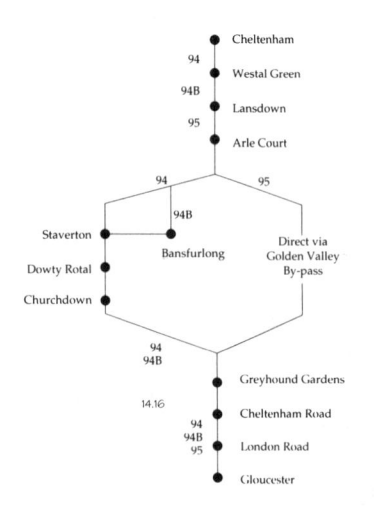

4. (a) Ben — Martin, John, Fergus; Fergus — Colin

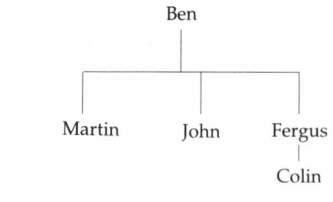

   (b) Martin → John → Fergus → Martin

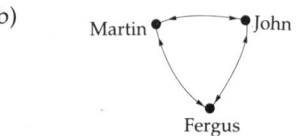

5. Angela → Ben → Caroline → Angela

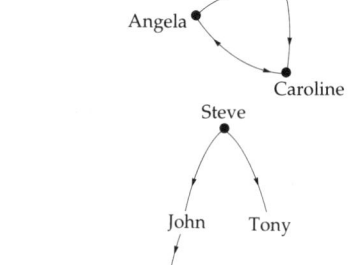

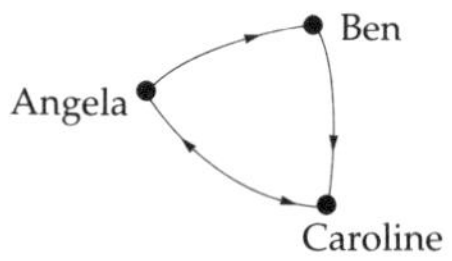

6. Steve ← John, Steve ← Tony, John ← David

7. Leckhampton beat Bourton – Home   2 pts
   Leckhampton beat Prestbury – Away   3 pts
   Winchcombe beat Prestbury – Home   2 pts
   Bourton beat Winchcombe – Home   2 pts

   Leckhampton 5 pts
   Winchcombe   2 pts
   Bourton   2 pts
   Prestbury   0 pts

15 Formulae and Equations

I am very worried about having to understand and use simple formulae or equations expressed in words. Formulae and equations mean algebra, don't they? I never understood what algebra was for. All those letters mean nothing to me.

I think most students feel like that about algebra but there is nothing to be frightened about. Algebra is a very old branch of mathematics. Almost all simple algebra was known by the fourth century AD. It's been around a long time.

I bet students have hated it for a long time too. All the words it uses are so strange. No one ever bothers to explain them. Mathematical mumbo jumbo! Even the word algebra is peculiar – sounds foreign to me.

*Algebra **is** foreign. It comes from the arabic al-Jabra which was the first word in the title of the first book about algebra. The book was written about AD825 by a mathematician called Al-Khowarizmi.*

Algebra

> I do not care who wrote the first book about algebra. I am expected to understand about formulae and equations and I am not happy about it. It has nothing to do with ordinary life. When do I use symbols and formulae in my life?

> Probably a lot more than you could ever believe. We are surrounded by symbols and abbreviations and formulae. I'll show a few that you will be familiar with. Symbols are part of our lives.

White: Demetrios Agnos
Black: Ali Mortazavi
Baroque International, February 1994
Sicilian defence

1	e4	c5
2	Nf3	e6
3	d4	cxd4
4	Nxd4	a6
5	Bd3	Nf6
6	0-0	Qe7
7	Qe2	d6
8	c4	g6
9	Nc3	Bg7
10	Be3	0-0
11	Rac1	b6
12	Rfd1	Bb7

GNVQ ▶

John Smith
2, Grange Road
Cambridge
CB 3 8DF

Drive from A to B

H_2O

Vauxhall Nova 1.2

BMW 318S

VOLVO 480ES

Cable Panel (Worked across 26 sts)
1st row: C4B, p6, C3B, C3F, p6, C4B.
2nd row: P4, k6, p6, k6, p4.
3rd row: K4, p5, C3B, k2, C3F, p5, k4.
4th row: P4, k5, p8, k5, p4.
5th row: C4B, p4, T3B, C4B, T3F, p4, C4B.
6th row: P4, k4, p2, k1, p4, k1, p2, k4, p4.
7th row: K4, p3, T3B, p1, k4, p1, T3F, p3, k4.
8th row: P4, k3, p2, k2, p4, k2, p2, k3, p4.
9th row: C4B, p2, T3B, p2, C4B, p2, T3F, p2, C4B.

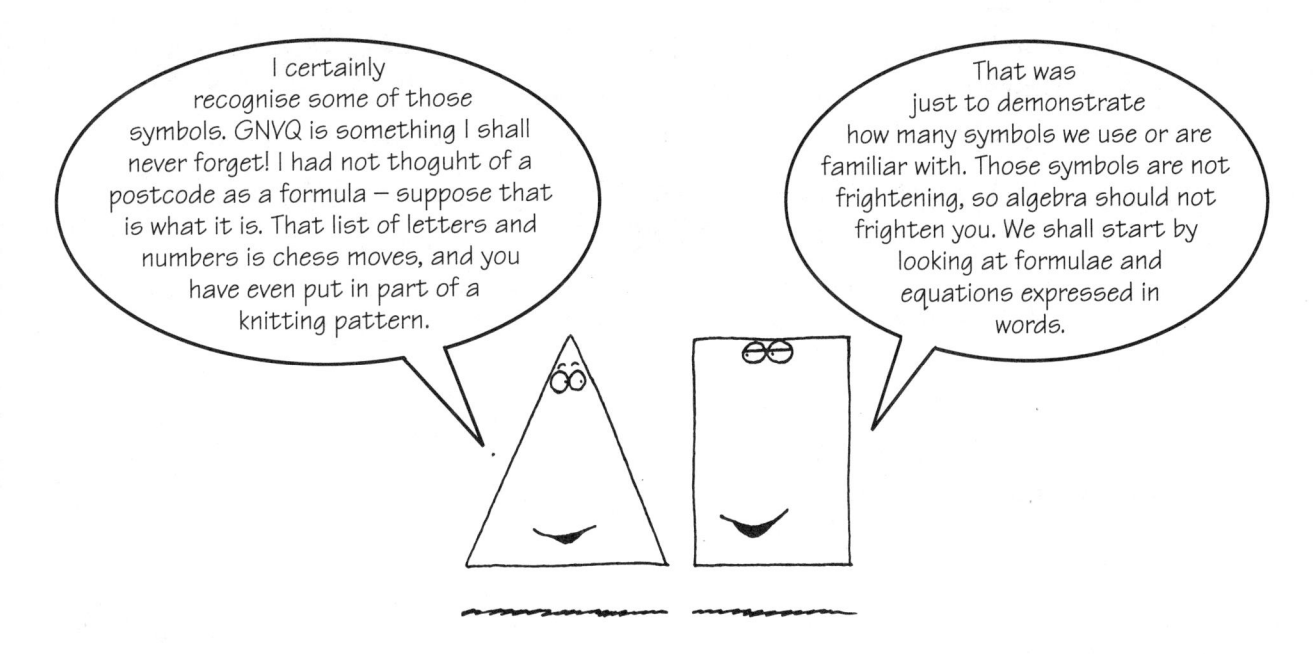

I certainly recognise some of those symbols. GNVQ is something I shall never forget! I had not thoguht of a postcode as a formula – suppose that is what it is. That list of letters and numbers is chess moves, and you have even put in part of a knitting pattern.

That was just to demonstrate how many symbols we use or are familiar with. Those symbols are not frightening, so algebra should not frighten you. We shall start by looking at formulae and equations expressed in words.

(a) Profit is equal to revenue minus cost.

(b) Balance is equal to income minus expenditure.

(c) Speed is equal to the distance travelled divided by the time taken on the journey.

(d) The distance travelled is equal to the speed multiplied by the time spent on the journey.

(e) The area of a rectangle is equal to the length multiplied by the width.

That is all so very boring and wordy. It is very difficult to remember all those sentences. You are just trying to convince me that algebra would make it all so much simpler.

I was trying to make a point. Formulae and equations expressed in words are difficult to remember and awkward to use. The same formulae and equations, written in the shorthand that is algebra, will be easier to learn and to use.

(a) Profit   = Revenue − Cost $P = R - C$
(b) Balance = Income − Expenditure $B = I - E$
(c) Speed   = Distance ÷ Time $S = D/T$
(d) Distance = Speed × Time $D = S \times T$
(e) Area   = Length × Width $A = L \times W$

In each case, a single letter symbol replaces the words. If you know what the symbol represents then the formula makes perfect sense.

> All the letters you have used in those formulae are just the first letters of the words they represent? Is that true all the time? I seem to remember letters like x, y and z in my algebra. They represented nothing in particular − just confused me.

> Where possible formulae will use letters relating to the subject. Sometimes this is not possible − I'll give you an example.

Here is a formula relating power and current.

$$\text{power} = \text{potential difference} \times \text{current}$$

When we write this in algebraic form we use P = power, E = potential difference, I = current. This is because of the conventions of electrical theory.

Our formula becomes

$$P = E \times I$$
$$= EI$$

> In algebra we leave out the multiplication sign. We say that 'multiplication is understood'.

Let's try putting numbers into each of the formulae.

(a) $P = R - C$ $R = £1500$ $C = £1250$
$P = 1500 - 1250$
$P = 250$ Profit = £250

(b) $B = I - E$ $I = £3650$ $E = £2550$
$B = 3650 - 2550$
$B = 1100$ Balance = £1100

(c) $S = D \div T$ $D = 112$ miles $T = 2.5$ h
$S = 112 \div 2.5$
$S = 44.8$ Speed $= 44.8$ mph

(d) $D = S \times T$ $S = 44.8$ mph $T = 2.5$ h
$D = 44.8 \times 2.5$
$D = 112$ Distance $= 112$ miles

(e) $A = L \times W$ $L = 21$ cm $W = 10$ cm
$A = 21 \times 10$
$A = 210$ Area $= 210$ cm^2

(f) $P = E \times I$ $E = 60$ volts $I = 220$ amperes
$= 60 \times 220$
$= 13\,200$ Power $= 13\,200$ watts

Putting numbers into a formula is not a problem. I have done that a lot when working out areas and volumes. I do not feel happy about making my own formula. That must be very difficult.

It is not so difficult. We will look at some examples to give you an idea of what to do. Then you can try some simple ones for yourself.

Example

Rectangle with sides 2 cm and 4 cm

The perimeter $= 4 + 2 + 4 + 2 = 2(4 + 2)$ cm

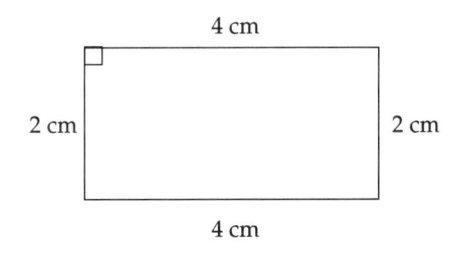

This rectangle has sides of length l and width w.
Perimeter $= l + w + l + w$
$= 2(l + w)$

The formula for the perimeter of a rectangle is

$$P = 2(l + w)$$

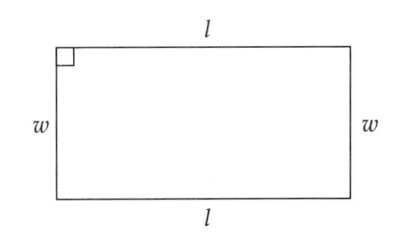

We can go on to construct a formula for the area of any rectangle.

Area of a rectangle is equal to the length × the width.

$$A = l \times w$$

$$A = lw$$

> **Remember:** In algebra we leave out the multiplication sign.

Example

Conversion between imperial and metric measures are easy when there is a simple formula to use.

$$1 \text{ mile} = 1.61 \text{ kilometres}$$

We use m for mile and k for kilometre.

$$k = 1.61\,m$$

This is a general formula that can be used for *any* value of m.

> It looks so easy when you do it. I'm sure I shall find it more difficult. Still, I shall never know until I try so can you give me some practice exercises?

Try some for yourself

1. Construct formulae for the following conversions:
   (a) kilometres to metres
   (b) metres to centimetres
   (c) grams to kilograms

2. Temperatures in degrees Celsius may be converted to degrees Fahrenheit by using the formula.
   $$F = \tfrac{9}{5}C + 32$$
   Convert to °F:
   (a) 100 °C   (b) 15 °C   (c) 0 °C

3. The conversion for pounds to Hong Kong dollars is
   £1 = 11.2 dollars
   (a) Construct a formula to convert from pounds to dollars.
   (b) Use your formula to convert 250 pounds to dollars.

4. Construct a formula for:
   (a) the perimeter of a square
   (b) the area of a square.

> I seem to manage reasonably well with constructing and using those formulae. What about equations? Are they the same as formulae?

> Formulae are special types of equations. Before we go on I think we need to look at some of the words associated with algebra. I have used words like formula, equation and symbol without explaining what I mean.

Words associated with algebra

Algebra: This is simply arithmetic written in a general form. Instead of using numbers we use symbols, usually letters, to represent different numbers.

Equation: This is a kind of mathematical statement where one part is made equal to the other part by use of the equals (=) sign.

Formula: A particular equation associated with a general result like the area of a circle.

Symbol: Any letter which is used to represent an unknown number.

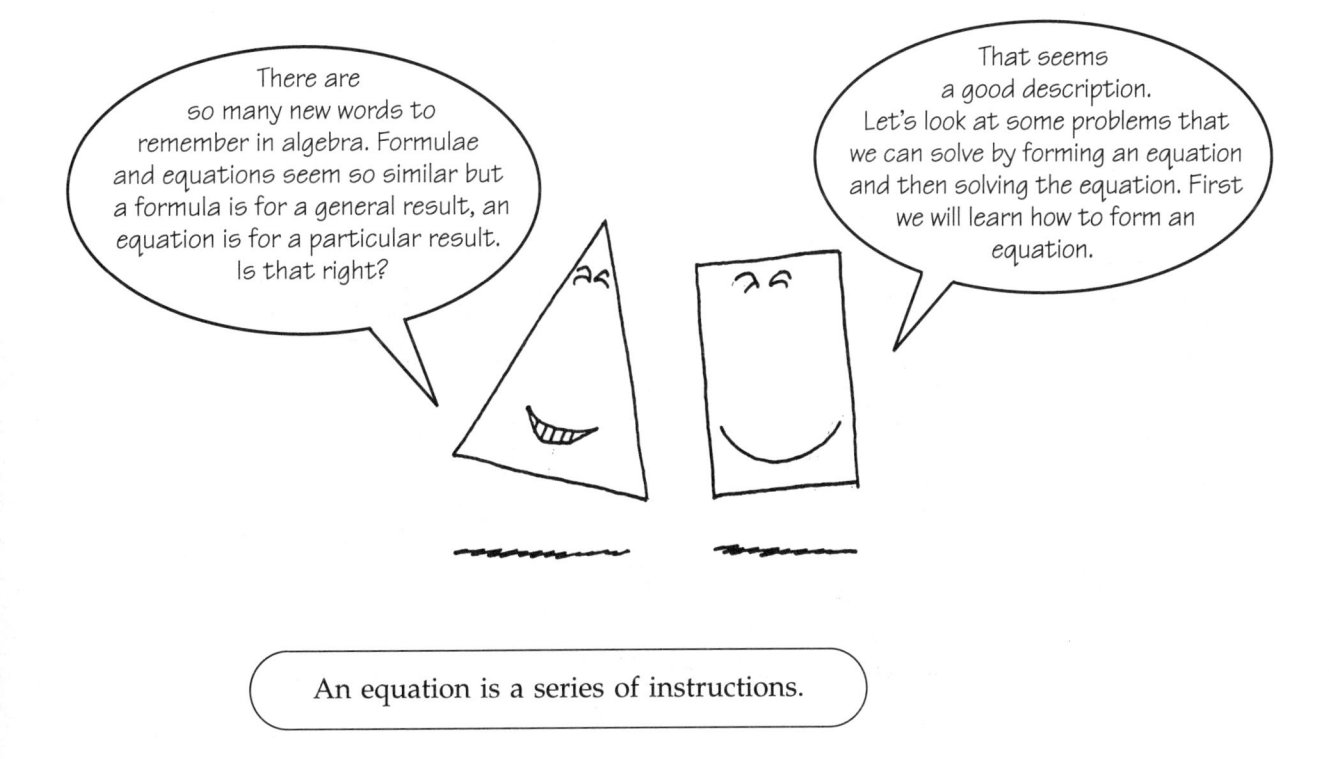

> There are so many new words to remember in algebra. Formulae and equations seem so similar but a formula is for a general result, an equation is for a particular result. Is that right?

> That seems a good description. Let's look at some problems that we can solve by forming an equation and then solving the equation. First we will learn how to form an equation.

An equation is a series of instructions.

245

Example

Grapefruit cost 18p each. I want to form an equation to help me work out the cost of buying any given number of grapefruit.

Find a symbol to represent grapefruit g
Multiply it by 18, the price of a grapefruit $18g$

The answer is the cost of buying g grapefruit C

$$C = 18g$$

If I buy 6 grapefruit, $g = 6$

$$C = 18 \times 6 = 108p = £1.08$$

I want you to help me with the next problem. It's very similar to the grapefruit one so you should be able to tell me what to do.

Oranges cost 10p each. We need to form an equation to allow us to work out the cost of buying any number of oranges.

Oranges: O — We need to find a symbol to represent oranges. Let's use the letter O.

$10O$ — We need to multiply by 10, the price of an orange.

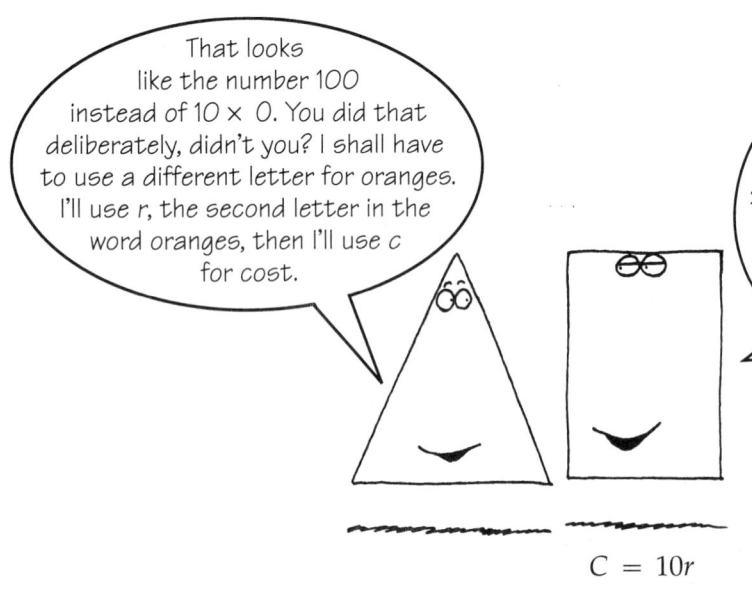

That looks like the number 100 instead of 10 × O. You did that deliberately, didn't you? I shall have to use a different letter for oranges. I'll use r, the second letter in the word oranges, then I'll use c for cost.

I did pick oranges so that you would realise that O is not a good letter to use in algebra. It looks like a zero and could be easily misread. Now you know why we cannot always use a letter that relates to the word we are representing. There is a good reason for everything in mathematics.

$$C = 10r$$

Example

The cost of hiring a small car is given as £25 per day plus a £50 deposit.

Form an equation to solve the cost of car hire.

Find a symbol for day	d
Multiply by £25	$25d$
Add £50	$25d + 50$

The answer is the cost c

$$c = 25d + 50$$

When the car is rented for 4 days

$$c = 25 \times 4 + 50$$
$$c = 100 + 50$$
$$c = £150$$

Example

My electricity bill is calculated like this.

Standing charge = £8.62
Each unit costs 7.37 pence

Form an equation to allow me to calculate my bill.

Find a symbol for a unit	u
Multiply by 7.37p	$7.37u$
Add the standing charge (in pence)	862

The answer is my bill B

$$B = 7.37u + 862$$

When I have used 720 units my bill will be

$$B = 7.37 \times 720 + 862$$
$$= 5306.4 + 862$$
$$= 6168.4 \text{ pence}$$
$$= £61.684$$

That was much more complicated. I would have forgotten to change the standing charge to pence and the answer would have been way out. I suppose you could have changed 7.37p to £0.0737?

Yes. It does not matter which way you convert as long as both figures are in the same units. That was a more complicated example but lots of bills are worked out that way so it is useful to know how to form the equation.

247

> Solving the equation is easy, it's putting the equation together that frightens me. Deciding on the symbols can also be a problem. I need practice.

> The most difficult part in solving a problem is choosing the right symbols and then forming the equation. Once the equation has been formed, the solution is quite easy. I'll give you some practice in choosing symbols and forming equations.

Try some for yourself

5. (a) A gallon of petrol costs £1.98. Form an equation to find the cost in pounds of any number of gallons of petrol.

   (b) Use your equation to find the cost of 10 gallons of petrol.

6. I want to hire a bouncy castle for the village fête. The cost per day is £35 and I shall have to make a deposit of £35.

   (a) Form an equation to calculate the bill.

   (b) Use your equation to find the cost of 3 days' hire.

7. My phone bill has a quarterly rental charge of £18.72. Each unit costs 3.62p.

   (a) Form an equation to work out my bill.

   (b) Use the equation to find the cost of 860 units.

   Remember: quarterly charge and unit charge must be in the same units.

What have you learnt about formulae and equations?

> The most important thing that I have learnt is that algebra is not frightening when you realise that it is just arithmetic using letters instead of numbers

- ✓ I can use a formula to solve problems by substituting numbers for letters.
- ✓ I can make a formula by writing down a list of instructions using a letter symbol to stand for the unknown quantity.
- ✓ I know that a formula is a particular type of equation.
- ✓ I know that an equation has two parts joined by an equals sign.
- ✓ I know how to form an equation when I am given information.
- ✓ I know how to solve my equations.

Further Questions

1. The relationship between power in watts (P), potential difference in volts (E) and current in amperes (I) is expressed in the formula.

$$P = EI$$

   (a) Use the formula to calculate the power required by a fire which draws 8 amperes froma 250 volt supply.

   (b) Use the formula to calculate the power used when an electric lamp, using 0.5 amperes is connected to a 240 volt supply.

2. Derive formulae for the following exchange rates:

   (a) pounds to lira   when £1 = 2450 lira
   (b) pounds to yen   when £1 = 154 yen
   (c) pounds to US dollars   when £1 = $1.445

3. The hire of a minibus is £65 per day plus a deposit of £75.

   (a) Form an equation to work out the bill.

   (b) Use your equation to find the cost of six days hire.

   The cost of hiring the minibus for one week is £325 plus a deposit of £75.

   (c) Would it be cheaper to hire the minibus for a whole week when I need it for only six days?

4. A couple wish to hire a travel cot for an extended holiday. A refundable deposit of £50 must be paid in advance. The hire charge is £9 per week.

   (a) Form an equation to calculate the total cost.

   (b) Use your equation to find the cost of hiring the cot for 6 weeks.

5. A carpet shampooing machine is hired by the half day. A refundable deposit of £30 is paid on collection of the machine. A charge of £7.50 per half day is quoted plus £9 for 2 litres of cleaning liquid.

   (a) Form an equation to calculate the final bill.

   (b) Use your equation to find the cost of hiring the machine for a day and a half.

6. The formula for converting from degrees Fahrenheit to degrees Celsius is given.

   $$C = \tfrac{5}{9}(F - 32)$$

   Use the formula to convert the following Fahrenheit temperatures to degrees Celsius.

   (a) 32°F

   (b) 40°F

   (c) 62°F

   (d) 81°F

7. The cost of driving lessons at a special student rate is:

   £11.50 per lesson
   £6.50 for the first lesson.

   (a) Form an equation to calculate the cost of a series of lessons.

   (b) Use your equation to find the cost of 6 lessons.

8. A mini-roundabout has a diameter of 1 m.

   (a) Use the formula for the circumference of a circle to find the circumference of the roundabout.

   (b) A large roundabout has a diameter of 38 m. Find the circumference.

9. A van is driven at night using headlights and tail lights. Each headlight bulb is 60 W and each tail light bulb 12 W. The van has an electrical system of 12 V.

I = current in amperes (A)
P = power in watts (W)
E = potential difference in volts (V).

Use the formula $I = \dfrac{P}{E}$ to find the current used when the van is being driven.

10. The formula for calculating the percentage profit is

$$P = \dfrac{S - C}{C} \times 100$$

P = profit percentage
S = selling price
C = cost price

If the selling price is £79.99 and the cost price is £42, calculate the profit percentage.

11. My telephone bill each quarter contains a rental charge of £19.54. The calls are charged at 3.78p per unit.

   (a) Form an equation to allow me to calculate my bill.

   (b) Use the equation to calculate my bills for the four quarters where the number of units were as follows:

   1st quarter   937
   2nd quarter   1430
   3rd quarter   1199
   4th quarter   898

   (c) Find my total bill for the year.

   (d) VAT at 17.5% is added to the bill.

   Calculate the VAT on my total bill.

   (e) Calculate the total + VAT.

12. The formula for calculating Simple Interest is

$$I = \dfrac{PRT}{100}$$

P = Principal, amount invested
R = Rate of interest
T = Term, length of the investment

If P = £2000
   R = 6.3%
   T = 4 years

Find the interest on the investment.

13. Gas bills are calculated with a standing daily charge of 10.10p. The gas used is charged at 1.47p per kWh (kilowatt hour).

   (a) Form an equation to allow me to calculate my bill for an unknown number of days *and* an unknown number of units.

(b) Use the equation to calculate bills for these readings.

1st quarter	102 days,	10 715 units
2nd quarter	94 days,	8700 units
3rd quarter	90 days,	3320 units
4th quarter	80 days,	3285 units

Solutions

1. (a) $m = 1000k$; $m =$ metres, $k =$ kilometres

(b) $c = 100m$; $c =$ centimetres, $m =$ metres

(c) $g = 1000k$; $g =$ grams, $k =$ kilograms

2. (a) $F = \frac{9}{5} \times 100 + 32 = 212°$

(b) $F = \frac{9}{5} \times 15 + 32 = 59°$

(c) $F = \frac{9}{5} \times 0 + 32 = 32°$

3. (a) $d = 11.2p$; d represents dollars, p represents pounds

(b) $d = 11.2 \times 250 = 2800$ dollars

4. (a) $4l$ $l =$ side of square

(b) $l \times l$ or l^2

5. (a) Use a symbol for gallon g
   Multiply by 1.98 $1.98g$
   Cost C
   $C = 1.98g$

(b) $C = 1.98 \times 10 = £19.80$

6. (a) $B = 35d - 35$ $B =$ bill, $d =$ days

(b) $B = 35 \times 3 - 35$
$= 105 - 35$
$= £70$

7. (a) $B = 18.72 + 0.0362\,u$ $B =$ bill, $u =$ units

(b) $B = 18.72 + 0.0362 \times 860$
$= 18.72 + 31.132$
$= £49.852$

Solutions to further questions

Chapter 1 Estimating and Calculating p. 9

1. £1
   £1
   £1
   £2
   £3
   £3
   £1
   £12

2. 250 + 70 + 130 + 110 = 560 miles

3. 2 × £500 + 1 × £300 = £(1000 + 300) = £1300
4. £40 + £25 + £30 = £95
5. £20 + £15 + £13 + £30 = £78
6. 2 × £190 + 2 × £140 = £660
7. 3 × 30p = 90p round to £1
   £1 + £4 + £4 + £1 + £1 + £1 = £12
8. 120 + 70 + 50 = 240 miles

Chapter 2 Directing Numbers p.29

1. (a) 13
   (b) −4
   (c) −7
   (d) 3
   (e) −12
   (f) 2
   (g) −1
   (h) −13
   (i) −20
   (j) 1
   (k) 4
   (l) 1
   (m) −7
   (n) 31
   (o) 19
   (p) 0
   (q) 33
   (r) 21

2. (a) −20
   (b) −20
   (c) +20
   (d) +20
   (e) −18
   (f) −2

   (g) $-\frac{1}{2}$ or −0.5
   (h) +5

3. (a) −3
   (b) 10
   (c) +6
   (d) +2
   (e) 1

   (i) +5
   (j) −2
   (f) 27
   (g) 10
   (h) 48
   (i) 16
   (j) 3

4. (a) −916
   (b) 15
   (c) 2 564 848
   (d) −3
   (e) −5
   (f) −180 588
   (g) 153 300
   (h) 627 450

Chapter 3 Fractions p.49

1. (a) 130   (b) 325   (c) 65

2. 92 + 23 + 14 = 129
   $$\text{Convalescence} = \frac{23}{129}$$

3. Fraction spent on services, etc. $= 1 - (\frac{3}{10} + \frac{1}{3})$
   $= 1 - \frac{19}{30}$
   $= \frac{11}{30}$

   £124.80 capital financing
   £138.67 staff costs
   £152.53 services, etc.

4. (a) $\frac{24}{432} = \frac{1}{18}$
   (b) 5 × 24 = 120
   $\frac{120}{432} = \frac{10}{36} = \frac{5}{18}$

5. Total number of shoes = 46 + 10 + 73 = 129
   (a) British $\frac{46}{129}$ (b) American $\frac{10}{129}$ (c) Continental $\frac{73}{129}$

6. (a) 32 × $\frac{3}{4}$ = 24
   35 × $\frac{3}{5}$ = 21
   45

   (b) $\frac{17}{32}$ upstairs + $\frac{25}{35}$ downstairs
   17 + 25 = 42
   Total seats = 32 + 35 = 67
   Fraction unoccupied = $\frac{42}{67}$

7. $\frac{1}{5} + \frac{1}{4} + \frac{3}{8} + \frac{1}{10} = \frac{8 + 10 + 15 + 4}{40} = \frac{37}{40}$
   $1 - \frac{37}{40} = \frac{3}{40}$

253

8.

	Normal price	Sale price
Sweatshirts	£16.99	£11.33
Grandad tops	£12.99	£ 8.66
Shorts	£ 9.99	£ 6.66
Jeans	£29.99	£19.99
Sweaters	£39.99	£26.66
Bodies	£21.99	£14.66
Leggings	£12.99	£ 8.66
Tunics	£14.99	£ 9.99

9. 4×8 hours at £3.72 $\quad$ = £119.04
 $+ \; 3$ hours at £3.72 $\times 1\frac{1}{2}$ $\quad$ = £ 16.74
 $+ \; 8$ hours at £3.72 $\times 2$ $\quad$ = £ 59.52
 $\qquad\qquad\qquad\qquad\qquad\quad$ £195.30

10. (a) $500 \times \frac{1}{25} = 20$ faulty

(b) $500 - 20 = 480$ satisfactory

(c) $1 - \frac{1}{25} = \frac{24}{25}$ satisfactory

Chapter 4 Decimals p.68

1. £11.80

2. £2 × 498 + £2 × 298.80 = £1593.60

3. 2 × £6.50 + 6 × £11.50 = £82

4. Total cost = £940.65   Change from £1000 = £59.35

5. £55.95

6. Total cost = £185.47   Change = £250 − 185.47 = £64.53

7. Total cost = £1135
 Irina and Sarah can afford the meal.

8. Hampton 2100
 Blue Ridge 1700
 Three Oaks 1900
 Orange 1600
 Pine Tree 2000
 Wingland 1800

9. (a) 6.4 min + 11.73 min = 18.13 min

(b) 4.08 miles + 6.37 miles = 10.45 miles

(c) 22.02 min

10. S. Gilbert: $\quad$ 126 ÷ 5 = 25.2
 T. Leigh: $\quad$ 72 ÷ 4 = 18
 A. Thompson: $\quad$ 84 ÷ 5 = 16.8
 C. Sainty: $\quad$ 98 ÷ 5 = 19.6
 D. Evans: $\quad$ 22 ÷ 3 = 7.3
 R. Fordham: $\quad$ 36 ÷ 3 = 12

11.

	Cambridge	Penzance	Newcastle
London	86.9	452.4	439.5

	Aberdeen	Holyhead
London	792.1	417

Distances in kilometres to 1 decimal place.

12. (a) (i) 2.79 $\quad$ (ii) 2.8

(b) (i) 0.05 $\quad$ (ii) 0.051

(c) (i) 8.17 $\quad$ (ii) 8.2

(d) (i) 0.05 $\quad$ (ii) 0.046

(e) (i) 126.09 $\quad$ (ii) 130

13. (a) £500 × 1.45 = 725 dollars

(b) 328 dollars remained

(c) £328 ÷ 1.49 = £220.13

14.

Volts	Amperes	Ohms
12	1.2	10
24	3.2	7.5
6	0.5	12
240	4.8	50
18	2.25	8
7.5	0.3	25
110	2.75	40
36	1.2	30

Chapter 5 Ratios, Scales and Percentages p.84

1.

Hotel	Child	Total
Days Inn	£239.20	£1674.40
Marriott	£405.60	£2163.20
Travelodge	£438.00	£2336

2. (a) £687.38

(b) £769.63

(c) £928.25

(d) £1163.25

3. £4.70

4. 17% + 42% = 59%
100 − 59% = 41% remaining ingredients
41% of 3.8 kg = 1.56 kg

5. 9.7%

6. (a) £52.50

(b) £39.38

7. 1320 ± 26 kg
1294 kg to 1346 kg

8. 3.3 tonnes

9. $\frac{65}{315} \times 100 = 20.63\%$

10. £3050 − 500 = £2550
£2550 + $2\frac{1}{2}$% commission = £2613.75
Commission = £63.75

11. $\frac{1}{4}$ paving : $\frac{3}{4}$ garden = 1 : 3

12. $\frac{2}{9} : \frac{3}{9} : \frac{4}{9}$ = £7292.89 : £10 939.33 : £14 585.78

13. 1 : 3 : 6   1 + 3 + 6 = 10 parts
$\frac{75}{10} = 7.5$
7.5 kg cement, 3 × 7.5 = 22.5 kg sand,
6 × 7.5 = 45 kg aggregate

14. 7 : 5   7 + 5 = 12 parts
$\frac{7}{12} \times 60 = 35$ men
$\frac{5}{12} \times 60 = 25$ women

15. Incoming calls : outgoing calls = 11 : 4
so incoming calls = $\frac{11}{4} \times 128$
= 352

16. 900 : 150
90 : 15
6 : 1

17. 300 000 : 150 000 : 50 000
30 : 15 : 5
6 : 3 : 1

18. (b) Liverpool to Newcastle   1.9 cm = 190 km

(c) Southampton to Manchester   2.8 cm = 280 km

(d) Norwich to Leeds   2.1 cm = 210 km

(e) Nottingham to Oxford   1.3 cm = 130 km

(f) Brighton to Plymouth   2.8 cm = 280 km

(g) Preston to Carlisle   1.2 cm = 120 km

(h) Aberdeen to Glasgow   1.9 cm = 190 km

(i) Nottingham to Middlesbrough 1.7 cm = 170 km

19. (a) 45 mm : 450 mm
Scale = 1 : 10

(b) Depth = 100 mm

(c) Width = 45 mm

20. (a) 500 × $\frac{3}{2}$ = 750 g
1.2 × $\frac{3}{2}$ = 1.8 litres

(b) 0.5 × $\frac{5}{2}$ = 1.25 kg
1.2 × $\frac{5}{2}$ = 3 litres

(c) 0.5 × 5 = 2.5 kg
1.2 × 5 = 6.0 litres

Chapter 6 Conversions p.99

1.

		Jan	Feb	Mar	Apr	May	June	July	Aug	Sept	Oct	Nov	Dec
Los Angeles	°F	65	66	65	67	69	72	75	77	77	74	70	66
	°C	18	19	18	19.5	20.5	22	24	26	26	23.5	21	19
Orlando	°F	72	74	78	82	88	90.5	90.5	90.5	90	82	77	72
	°C	22	23	25.5	28	31	33	33	33	32	28	25	22

2. (a) (i) 40   (ii) $38\frac{1}{2}$ (iii) $36\frac{1}{2}$

(b) (i) $3\frac{1}{2}$ (ii) 5   (iii) $5\frac{1}{2}$

3. (a) 15 mpg = 24.15 kpg

(b) 20 mpg = 32.2 kpg

(c) 30 mpg = 48.3 kpg

(d) 35 mpg = 56.35 kpg

4. (a) (i) 40   (ii) 44
   (b) (i) 12   (ii) 18

5. 9 oz = 255.15 g
   6 oz = 170.10 g
   4 oz = 113.40 g
   7 oz = 198.45 g

6. 175 g = 7 oz
   175 g = 7 oz
   225 g = 9 oz
   75 g = 3 oz
   50 g = 2 oz
   25 g = 1 oz
   100 g = 4 oz
   225 g = 9 oz
   450 g = 18 oz

7. (a) 157.5 mA
   (b) 5.7 V

8. (b) (i) $\frac{1}{2}$ mile = 0.8 km
     (ii) 3.5 miles = 5.64 km
     (iii) 1.6 miles = 2.58 km
     (iv) 10 miles = 16.1 km
     (v) 2 miles = 3.22 km

9. £   0   10   20
   Mark 0   25   50

(a) 62.5 Marks
(b) 250 Marks
(c) £16
(d) £40

Chapter 7   Surveys and Questionnaires   p.111

1. Which bank do you use?

   Lloyds ☐
   Midland ☐
   Nat West ☐
   Barclays ☐
   Other ☐

If other, give the name of your bank/building society ....

Which banking facilities do you use?

   Withdrawal ☐
   Balance enquiry ☐
   Statement request ☐
   Other ☐

If other, give details of the facilities you use ......

How often do you visit your bank

   Daily ☐
   Twice a week ☐
   Four times a week ☐
   Weekly ☐
   Other ☐

If other, give details ....

What type of account do you have?

   Young Persons ☐
   Current ☐
   Deposit ☐
   Other ☐

If other, give details ....

2. (a) If the young people have left, he can not ask the question.

  (b) The question could be worded to find out, from current customers, why people have left.

     The manager could ask for opinions of the facility and in this way learn what people like/dislike.

3. (a) This question assumes all students arrived late to begin with. It is also too vague, there is no time scale.

  (b) Does this mean immediate family, e.g. parents, brothers and sisters, or extended family?

  (c) What do you count as a room?

  (d) This question is too vague and the respondent has to know about healthy habits, etc. A better question would require the respondent to tick an appropriate number of boxes, given a wide range of options.

Chapter 8 Tabulating your Information p.120

1. Category C:

Rooms	Price	Swimming Pool
252	49	✓
147	49	✓
122	54	✓
130	43	

Category G:

Rooms	Price	Swimming Pool
74	79	
104	74	

Category H:

Rooms	Price	Swimming Pool
42	49	
57	49	
118	54	
52	54	✓
67	43	
87	58	
53	49	
34	64	
55	64	✓
40	58	
58	54	
20	58	
63	74	
37	59	
47	54	
29	54	

Category P:

Rooms	Price	Swimming Pool
192	49	✓
136	45	
184	45	
122	45	✓
88	45	✓
60	45	

2. (a)
Exeter	0600	0659	0837	1131	1437
Taunton	0627	0727	0905	1155	1505
Reading	0802	0857	1037	1310	1633
Exeter	1529	1637	1729	1936	
Taunton	1553	1701	1757	2005	
Reading	1711	1829	1911	2134	

(b)
Plymouth	0600	0735	0835	1135	1535
Paddington	0930	1110	1150	1453	1900
Plymouth	1835	Sat.	11.35		
Paddington	2210	only	14.51		

Chapter 9 Illustrating Information p.145

1. Facilities for visitors at Royal Botanical Gardens

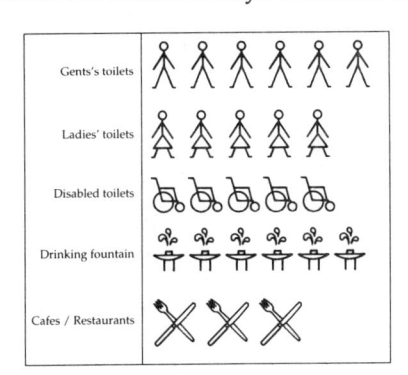

2. Sunday newspaper readership by year

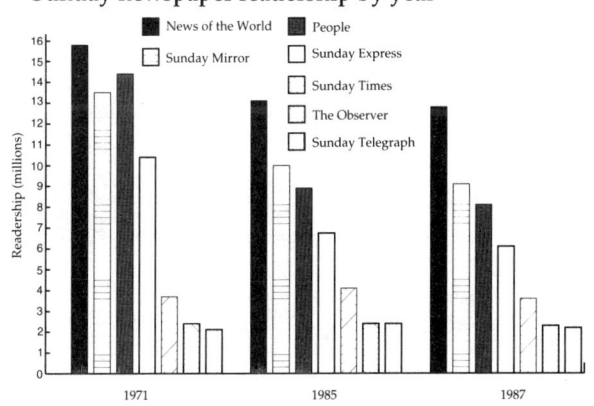

257

Sunday newspaper readership by paper

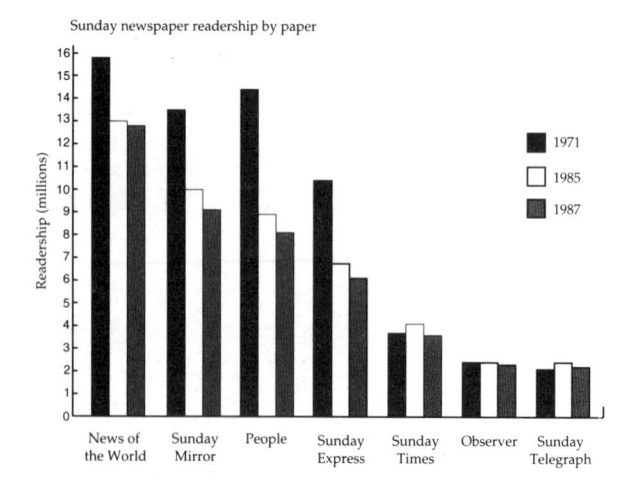

[Bar chart showing readership in millions for 1971, 1985, 1987 across News of the World, Sunday Mirror, People, Sunday Express, Sunday Times, Observer, Sunday Telegraph]

3. The numbers taking private medical insurance rose slowly between 1956 and 1978. From 1978 to 1982 the figures rose rapidly, they continue to rise but at a slower rate. As the National Health Service waiting lists increased, more people are turning to private medical care.

4. Suitable graphs would be pie charts or bar charts.

5. The greatest increase in enrolment in 1955–67 was in Japan with figures almost double those of the UK. The smallest increase was in the UK.

6. Suitable charts would be pie charts or bar charts.

   Between 1971 and 1981 car ownership in Japan doubled. The growth in all the countries was greater in the period between 1971 and 1981. Growth was slower between 1981 and 1986 – this is a 5-year period so projected figures for 1991 would indicate similar growth patterns in the UK and US, with a smaller growth in Japan. The USA is the largest car producing and car owning country.

7. Radio has its largest audiences before 12 noon – people listened to radio while working in the home. Television viewing rises rapidly from 6p.m. to 10p.m. – the peak viewing hours. The audience figures fall rapidly from 11p.m. and by 4a.m., they are minimal.

8. Readership of Women's Magazines-by title

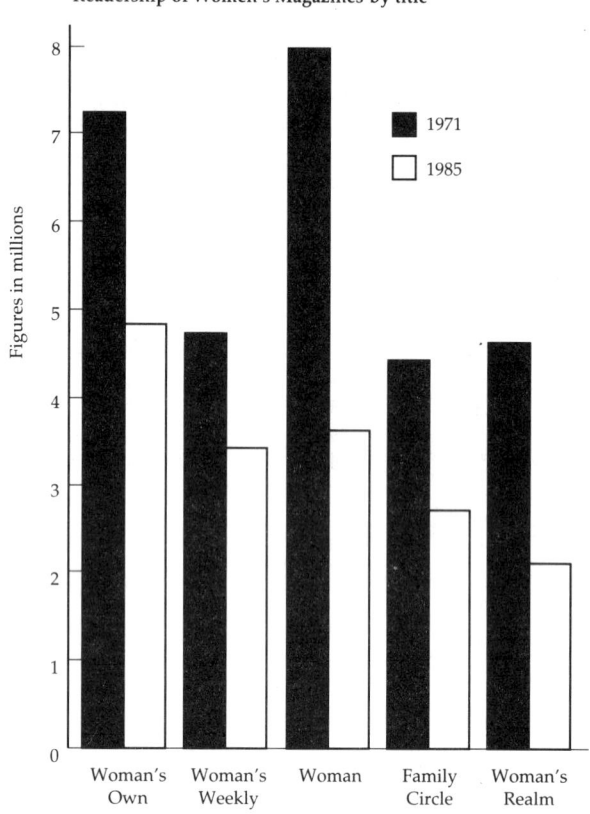

[Bar chart showing figures in millions for 1971 and 1985 across Woman's Own, Woman's Weekly, Woman, Family Circle, Woman's Realm]

This data can also be illustrated by year.

The readership of all women's magazines dropped in the period from 1971 to 1985. The most dramatic figures are those for Woman magazine where the readership fell by over 50%. New magazines aimed at the younger career woman were launched during this period taking readers from the older magazines.

9. Between 1983 and 1986 ownership of video cassette recorders has risen significantly. Families with 1, 2 or 3 children are most likely to own a VCR. Adults living alone are the least likely owners. Peer pressure and changing patterns of domestic life may be factors in the rise in use.

10. Compound or component bar charts or pie charts are suitable methods of illustration.

11. The largest television audience are those people over 65 who are retired and at home all the time. Children view fewer hours as there are fewer suitable programmes for the age group.
Children view fewer hours as there are fewer suitable programmes for the age group.

12. Compound bar charts are the most suitable. Vaccinations for poliomyelitis and measles show a steady rise over the 16-year period. Vaccinations for whooping cough dropped by more than 50% in 1976 when there was widespread anxiety about the side effects of the vaccine. Figures rose again from 1981 and had almost reached the 1971 level by 1987.

13. 43% of 20–4 year olds are in HE in the US. Only 14% of 20–4 year olds are in HE in the UK. The only countries with lower figures are Spain and Germany.

14. Day – bar chart
Time – pie chart
Role – pie chart/bar chart
Ages – bar chart

Chapter 10 Measuring Information p.166

1.

£43	11	2
£45	1111	5
£49	1111 1	6
£54	1111 1	6
£58	111	3
£59	1	1
£64	11	2
£74	11	2
£79	1	1
	Total	28

(a) Mean price = £54.18

(b) Median = $\dfrac{14\text{th} + 15\text{th}}{2}$

14th and 15th values are in the £54 band width

so median = $\dfrac{54 + 54}{2}$ = £54

(c) There are two modes. This is quite acceptable. The two most popular prices are £49 and £54.

(d) The range = 79 − 43 = £36

2. (a) Median = $\dfrac{15\text{th} + 16\text{th}}{2} = \dfrac{50 + 50}{2}$ = 50 bulbs

(b) Mean =
$\dfrac{1 \times 10 + 3 \times 15 + 5 \times 25 + 12 \times 50 + 9 \times 100}{30}$

= $\dfrac{1680}{30}$

= 56 bulbs

(c) Mode = 50 bulbs

3. (a) Modal salary = £8500

(b)

Number	Annual salary
1	£4250
4	£6550
9	£8500
2	£9500
1	£19 500
1	£19 500
1	£30 000
1	£45 000

Median = $\dfrac{10\text{th} + 11\text{th}}{2} = \dfrac{£8500 + 8500}{2}$ = £8500

(c) Mean = $\dfrac{£239\,950}{20}$ = £11 997.50

(d) Salary range = £45 000 − £4250
= £40 750

4. Mean =
$\dfrac{207 + 182 + 207 + 218 + 203 + 208 + 212}{7} = \dfrac{1437}{7}$

= 205.29 to 2 d.p.

Median: 182 203 207 207 208 212 218
                              ↑
                           Middle

Median = 207

Mode = 207

Chapter 11 Probability p.176

1. Total number of students = 40

(a) Number of Size 9 = 15
Probability of 9 = $\frac{15}{40} = \frac{3}{8}$

(b) Number of size 11 + Size 12 = 3 + 3 = 6
Probability of 11 or 12 = $\frac{6}{40} = \frac{3}{20}$

2. Probability of win = $\frac{100}{1\,000\,000\,000} = \frac{1}{10\,000\,000}$

3.

Coin 1	Coin 2	Coin 3	Probability
H	H	H	$\frac{1}{8}$
H	H	T	$\frac{1}{8}$
H	T	H	$\frac{1}{8}$
T	H	H	$\frac{1}{8}$
T	T	H	$\frac{1}{8}$
T	H	T	$\frac{1}{8}$
H	T	T	$\frac{1}{8}$
T	T	T	$\frac{1}{8}$

(a) Probability of 3 heads = $\frac{1}{8}$

(b) Probability of 2 heads and 1 tail = $\frac{1}{8} + \frac{1}{8} + \frac{1}{8} = \frac{3}{8}$

(c) Probability of 3 tails = $\frac{1}{8}$

4.

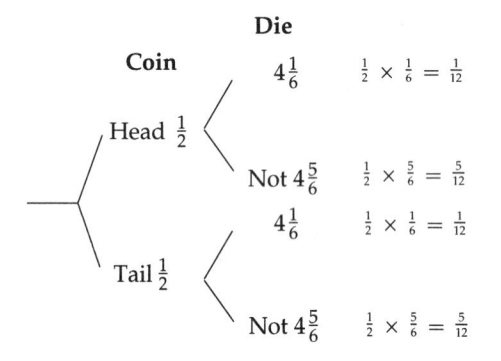

```
        Die
Coin
        4  1/6      1/2 × 1/6 = 1/12
Head 1/2
        Not 4  5/6  1/2 × 5/6 = 5/12

        4  1/6      1/2 × 1/6 = 1/12
Tail 1/2
        Not 4  5/6  1/2 × 5/6 = 5/12
```

Probability of head then 4 = $\frac{1}{12}$

5.

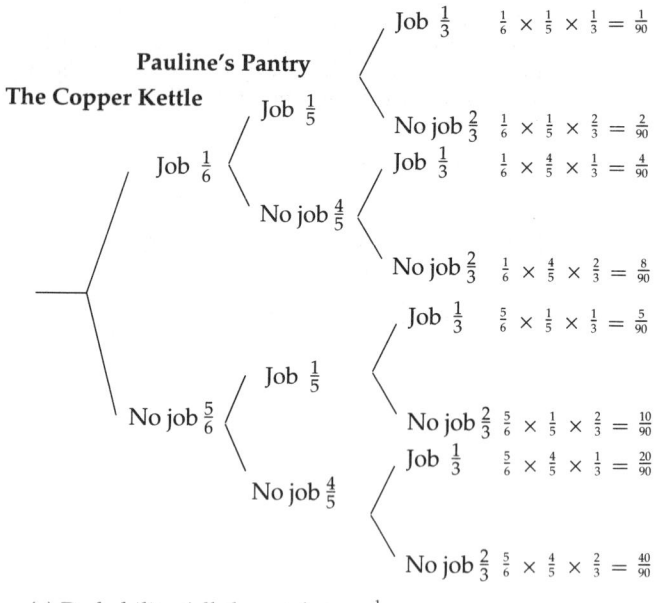

```
                              Chez Charles
                    Pauline's Pantry   Job 1/3    1/6 × 1/5 × 1/3 = 1/90
The Copper Kettle   Job 1/5
        Job 1/6                        No job 2/3  1/6 × 1/5 × 2/3 = 2/90
                                       Job 1/3    1/6 × 4/5 × 1/3 = 4/90
                    No job 4/5
                                       No job 2/3  1/6 × 4/5 × 2/3 = 8/90
                                       Job 1/3    5/6 × 1/5 × 1/3 = 5/90
                    Job 1/5
        No job 5/6                     No job 2/3  5/6 × 1/5 × 2/3 = 10/90
                                       Job 1/3    5/6 × 4/5 × 1/3 = 20/90
                    No job 4/5
                                       No job 2/3  5/6 × 4/5 × 2/3 = 40/90
```

(a) Probability (all three jobs) = $\frac{1}{90}$

(b) Probability (Copper kettle only) = $\frac{8}{90} = \frac{4}{45}$

(c) Probability (No job) = $\frac{40}{90} = \frac{4}{9}$

(d) Probability (One job only) = $\frac{8}{90} + \frac{10}{90} + \frac{20}{90} = \frac{38}{90} = \frac{19}{45}$

6.

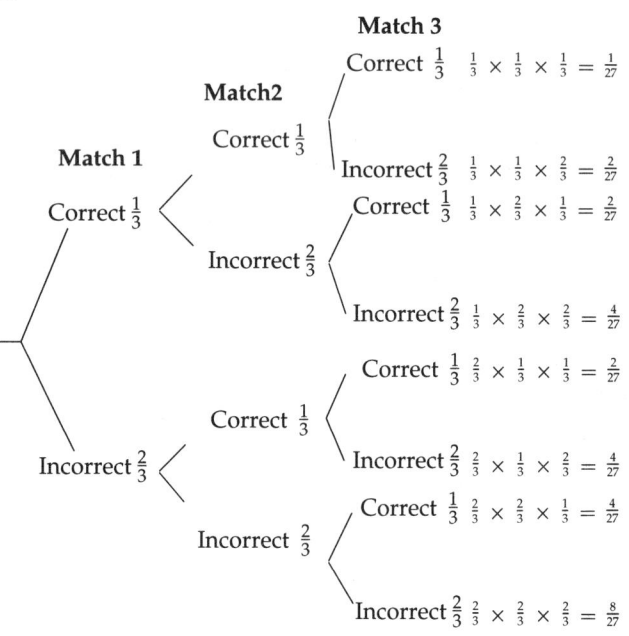

```
                              Match 3
                    Match 2   Correct 1/3   1/3 × 1/3 × 1/3 = 1/27
        Match 1     Correct 1/3
        Correct 1/3            Incorrect 2/3  1/3 × 1/3 × 2/3 = 2/27
                              Correct 1/3   1/3 × 2/3 × 1/3 = 2/27
                    Incorrect 2/3
                              Incorrect 2/3  1/3 × 2/3 × 2/3 = 4/27

                              Correct 1/3   2/3 × 1/3 × 1/3 = 2/27
                    Correct 1/3
        Incorrect 2/3          Incorrect 2/3  2/3 × 1/3 × 2/3 = 4/27
                              Correct 1/3   2/3 × 2/3 × 1/3 = 4/27
                    Incorrect 2/3
                              Incorrect 2/3  2/3 × 2/3 × 2/3 = 8/27
```

Probability of 3 correct forecasts = $\frac{1}{27}$

Probability of at least 2 correct = $\frac{1}{27} + \frac{2}{27} + \frac{2}{27} + \frac{2}{27} = \frac{7}{27}$

12 Perimeters and Areas p.203

1. (a) Area of room = Area of rectangle
2.5×3.4 m² − area of rectangle 1.5×0.3 m²
$= (8.5 - 0.45)$ m² $= 8.05$ m²

(b) Area of carpet = 2.7×3.6 m² = 9.72 m²
Area wasted = $(9.72 - 8.05)$ m² = 1.67 m²

2. (a) Area of carpet = 2.5×2.5 m² = 6.25 m²

(b) Area of cabinet = 0.45×0.6 m² = 0.27 m²
Area of 2 desks = $2(1.5 \times 0.75)$ m² = 2.25 m²
Area of bookcase = 0.75×0.3 m² = 0.225 m²
Area of table = 1×0.6 m² = 0.6 m²
Total Area of furniture = 3.345 m²

(c) Area of unoccupied floor space =
$(6.25 - 3.345)$ m² = 2.905 m²

3. Area of patio = 2×3 m² = 6 m²
Area of each slab = 0.6×0.6 m² = 0.36 m²
Number of slabs = $\dfrac{\text{Area of patio}}{\text{Area of slab}} = \dfrac{6}{0.36}$
$= 16.67$ to 2 d.p.
Must buy a whole number of slabs so we need 17 slabs.

4. (a) Perimeter of court = $2(30.5 + 15.25)$ m
$= 2 \times 45.75$ m = 91.5 m

(b) Area of court = 30.5×15.25 m²
$= 465.125$ m²

(c) Area of centre third = $465.125 \div 3$
$= 155.04$ m² to 2 d.p.

(d) Area of centre circle = $\pi r^2 = \pi \times (\frac{0.9}{2})^2$ m²
$= 0.64$ m² to 2 d.p.

(e) Area of goal circle = $\frac{1}{2}\pi r^2 = \frac{1}{2}\pi \times 4.9^2$ m²
$= 37.7$ m² to 2 d.p.

(f) Circumference of centre circle = πd
$= \pi \times 0.9$ m = 2.83 m to 2 d.p.

5. Perimeter = $2(120 + 92)$ m = 2×212 m = 424 m
Area = 120×92 m² = $11\,040$ m²

6. (a) A5 $2(105 + 148.5)$ mm = 507 mm
A4 $2(210 + 297)$ mm = 1014 mm
A3 $2(420 + 297)$ mm = 1434 mm

(b) A5 105×148.5 mm² = 15549.25 mm²
A4 210×297 mm² = $62\,370$ mm²
A3 420×297 mm² = $124\,740$ mm²

7. (a) Area = 90×260 cm² = $23\,400$ cm²

(b) Area = 75 cm $\times 4$ m = 0.75 m $\times 4$ m = 3 m²

8. (a) $30 \div 2.5 = 12$
12 beds along each wall = 24 beds

(b) Floor area = 30×24 m²
$= 720$ m²

9. (a) Circumference of semicircle of diameter 12 m
$= \frac{1}{2}\pi d = 6 \times \pi = 18.45$ m to 2 d.p.
Perimeter of track = $2 \times 18.45 + 2 \times 200$ m
$= 437$ m

(b) Area = area 2 semicircles radius 6 m
+ area rectangle 12×200 m²
Area semicircle = $\frac{1}{2}\pi r^2 = \frac{1}{2}\pi \times 6^2$
$= 56.55$ m² to 2 d.p.
Area of 2 semicircles = 2×56.55 = 113.1 m²
Area of rectangle 12×200 m² = 2400 m²
Area of track = 2513.1 m²

10. Circumference of table = $2\pi r = 2\pi \times 100$ cm
$= 628.3$ cm

(a) Allow 65 cm per customer
$628.3 \div 65 = 9.67$ customers
$= 9$ customers

(b) Allow 62 cm per customer
$628.3 \div 62 = 10.1$ customers
$= 10$ customers

11. Area of room = 3×4 m² = 12 m²
Area of tile = 0.25×0.25 m² = 0.0625 m²

(a) Number of tiles = $\dfrac{\text{Area of room}}{\text{Area of tiles}} = \dfrac{12}{0.0625}$

$= 192$ tiles

(b) Boxes of 5 = $192 \div 5 = 38.4$ boxes
Must buy 39 boxes.

12. (a) $2.5 + 1.5 + 2 + 2.5 + 2$ km = 10.5 km

(b) $1.5 + 2 + 4.5 + 2.5 + 1.5$ km = 12 km

(c) $5 + 3.5 + 4 + 5 + 3 + 4 + 2$ km = 26.5 km

13. (a) Area = $\pi r^2 = \pi \times 22^2$ cm² = 1520.53 cm²

(b) Scoring area = $\pi r^2 = \pi \times 17^2$ cm² = 907.92 cm²

(c) Area of centre circle = $\pi r^2 = \pi \times 1.8^2$ cm²
$= 10.18$ cm²

(d) Non-scoring area = Area of board − scoring area
$= 1520.53 - 907.92$ cm²
$= 612.61$ cm²

14. 32 cm $\times 22$ cm = 704 cm² to be coated

15. (a) Area of court = 26×14 m² = 364 m²

(b) Perimeter = $2(26 + 14)$ m = 2×40 m = 80 m

(c) Length of rope = $2(26 + 2 + 14 + 2)$ m
$= 2 \times 44$ m = 88 m

16. Area of side = $\frac{1}{2}(1 + 6) \times 23$ m² = $\frac{7}{2} \times 23$ m²
$= 80.5$ m²

17. Area of side = $\frac{1}{2}(3 + 5) \times 7.5$ m² = 4×7.5 m²
$= 30$ m²

261

Chapter 13 Volumes p.223

1. (a) $18 \times 30 \times 8 \, m^3 = 4320 \, m^3$
(b) $0.9 \times 0.9 \times 2.4 \, m^3 = 1.944 \, m^3$
(c) $1.8 \times 1.8 \times 2.2 \, m^3 = 7.128 \, m^3$

2. (a) Volume of pool $= 24 \times 19 \times 2 \, m^3 = 9.12 \, m^3$
(b) Amount of water $= 912 \, l$

3. $d = 6 \, cm \quad r = 3 \, cm \quad h = 10 \, cm$
$V = \pi r^2 h = \pi \times 3^2 \times 10 \, cm^3 = 282.74 \, cm^3$
Capacity $= 282 \, ml = 0.282 \, l$

4. $d = 6.4 \, cm \quad r = 3.2 \, cm \quad h = 12 \, cm$
(a) $V = \pi r^2 h = \pi \times 3.2^2 \times 12 \, cm^3 = 386.04 \, cm^3$
(b) 386 ml
(c) $386 - 325 \, ml = 61 \, ml$
The can is not full.

5. $V = l \times w \times h$
$53\,000 = 72 \times 48 \times h \quad h = \frac{53\,000}{72 \times 48}$
$h = 15.33 \, m$

6. (a) $r = 13 \, cm \quad h = 30 \, cm$
$V = \pi r^2 h$
$= \pi \times 13^2 \times 30 \, cm^3 = 15\,927.87 \, cm^3$
(b) Capacity $= 15.928$ litres $= 15.93$ litres to 2 d.p.

7. $d = 5.0 \, cm \quad r = 2.5 \, cm \quad h = 125 \, cm$
$V = \pi r^2 h = \pi \times 25^2 \times 125 \, cm^2$
$= 245\,436.93 \, cm^3$

8. $V = l \times w \times h$
$7005 = l \times 33 \times 9.7$
$l = \frac{7005}{33 \times 9.7} \, cm = 21.88 \, cm$

9. $V = \frac{1}{2} \, base \times height \times length$
$= \frac{1}{2} \times 120 \times 90 \times 190 \, cm^3$
$= 1\,026\,000 \, cm^3$

10. $V = \frac{1}{2} \times 122 \times 100 \times 190 \, cm^3 = 1\,159\,000 \, cm^3$

11. (a)

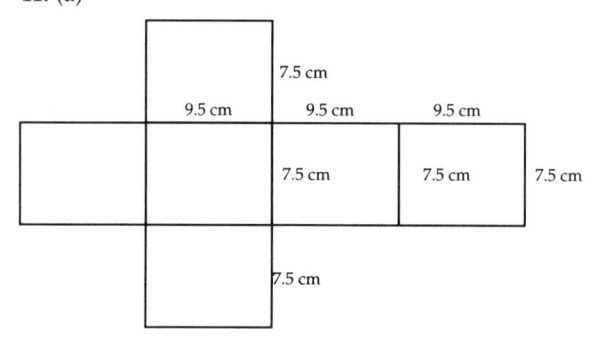

(b) $4 \times 9.5 \times 7.5 \, cm^2 \quad = 285 \, cm^2$
$2 \times 7.5 \times 7.5 \, cm^2 \quad = 112.5 \, cm^2$
Total area $\quad = \overline{397.5} \, cm^2$

12. (a)

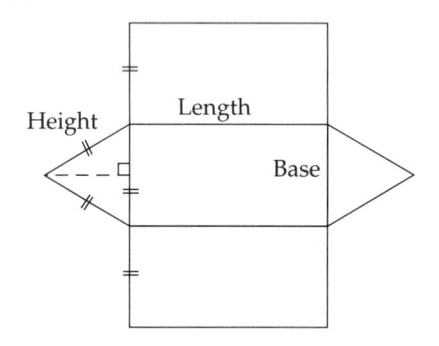

3 nets similar to this with appropriate measurements.

(b) Fun
$2 \times 0.9 \times 8 \quad = 14.4$
$2 \times 0.9 \times 1 \quad = 1.8$
$2 \times 1 \times 8 \quad = 16$
Total paper needed $= \overline{32.2} \, cm^2$

Standard
$2 \times 1.7 \times 16 \quad = 54.4$
$2 \times 1.7 \times 2 \quad = 6.8$
$2 \times 2 \times 16 \quad = 64$
Total paper needed $= \overline{125.2} \, cm^2$

Giant
$2 \times 3.5 \times 32 \quad = 224$
$2 \times 3.5 \times 4 \quad = 28$
$2 \times 4 \times 32 \quad = 256$
Total paper needed $= \overline{508} \, cm^3$

Chapter 14   Networks – what are they?   p.236

1. (a)

[Map showing cities Bergen, Haarlem, Amsterdam, The Hague, Rotterdam with distances: Bergen–Haarlem 31 km, Bergen–Amsterdam 36 km, Bergen–The Hague 72 km, Bergen–Rotterdam 85 km, Haarlem–Amsterdam 18 km, Haarlem–The Hague 42 km, Haarlem–Rotterdam 53 km, Amsterdam–The Hague 52 km, Amsterdam–Rotterdam 58 km, The Hague–Rotterdam 22 km]

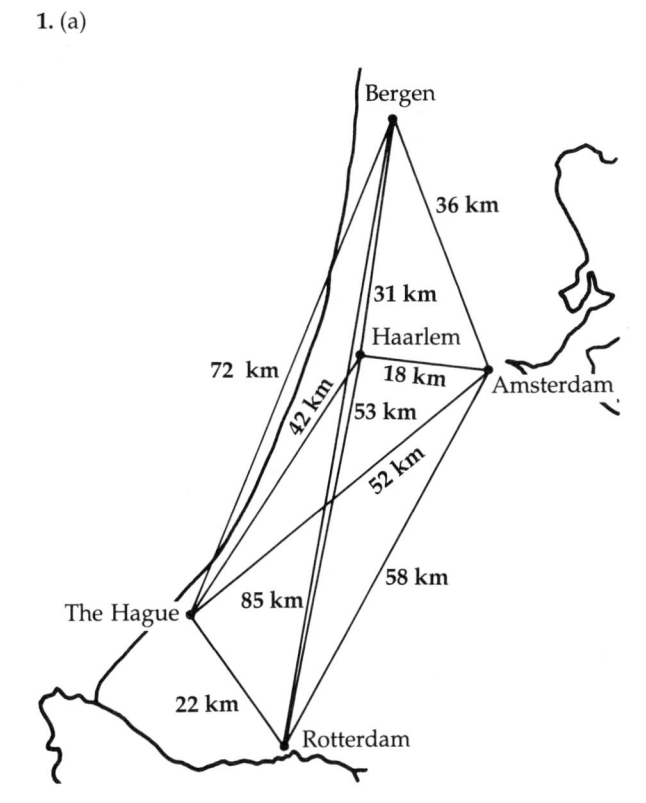

Scale: 15 km, 30 km, 45 km, 60 km

Rotterdam –	The Hague	22 km
	Haarlem	53 km
	Amsterdam	58 km
	Bergen	85 km
Amsterdam –	Bergen	36 km
	The Hague	52 km
	Rotterdam	58 km
	Haarlem	18 km
Bergen –	Amsterdam	36 km
	Haarlem	31 km
	Rotterdam	85 km
	The Hague	72 km
The Hague –	Rotterdam	22 km
	Amsterdam	52 km
	Haarlem	42 km
	Bergen	72 km
Haarlem –	Bergen	31 km
	Amsterdam	18 km
	The Hague	42 km
	Rotterdam	53 km

(b) The Hague – Haarlem – Bergen – Rotterdam – Amsterdam = 216 km
The Hague – Rotterdam – Bergen – Haarlem – Amsterdam = 156 km
The Hague – Bergen – Rotterdam – Haarlem – Amsterdam = 228 km
The Hague – Bergen – Haarlem – Rotterdam – Amsterdam = 214 km
The Hague – Haarlem – Rotterdam – Bergen – Amsterdam = 216 km
The Hague – Rotterdam – Haarlem – Bergen – Amsterdam = 142 km

The longest route is 228 km.

(c)

[Network diagram of the five cities with edges labelled: Bergen–Haarlem 31, Bergen–Amsterdam 36, Haarlem–Amsterdam 18, Haarlem–The Hague 42, Haarlem–Rotterdam 53, Bergen–The Hague 72, Bergen–Rotterdam 85, Amsterdam–The Hague 52, Amsterdam–Rotterdam 58, The Hague–Rotterdam 22]

2.

[Organisation chart:
Manager → Sales Manager, Service Manager, Parts Manager, Forecourt Manageress
Sales Manager → Salesman, Salesman
Service Manager → Bodyshop Manager, Foreman Mechanic (→ 4 mechanics); Bodyshop Manager → 2 panel breakers
Parts Manager → Assistant, Assistant
Forecourt Manageress → Cashier, Cashier]

3.

[Organisation chart:
Chief Executive → Head of Property Co., Head of Construction, Head of Engineering, Head of Development, Head of Services]

263

4. (a) (i)

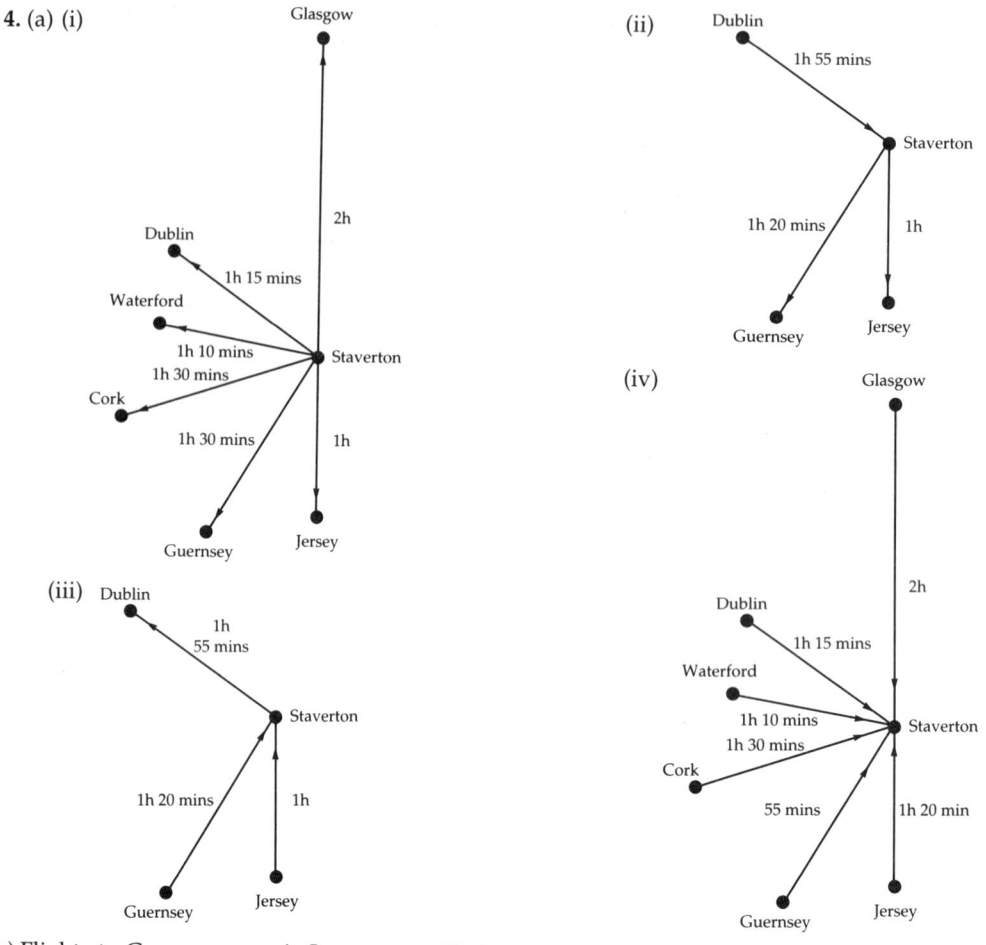

[Graph with Staverton as central node, connected to: Glasgow (2h, outgoing), Dublin (1h 15 mins, incoming), Waterford (1h 10 mins, outgoing), Cork (1h 30 mins, outgoing), Guernsey (1h 30 mins, outgoing), Jersey (1h, outgoing)]

(ii) [Graph: Dublin → Staverton (1h 55 mins); Staverton → Guernsey (1h 20 mins); Staverton → Jersey (1h)]

(iii) [Graph: Staverton → Dublin (1h 55 mins); Guernsey → Staverton (1h 20 mins); Jersey → Staverton (1h)]

(iv) [Graph with Staverton as central node: Glasgow (2h, incoming), Dublin → Staverton (1h 15 mins), Waterford → Staverton (1h 10 mins), Cork → Staverton (1h 30 mins), Guernsey → Staverton (55 mins), Jersey → Staverton (1h 20 min)]

(b) Flights to Guernsey are via Jersey.   Flights to Cork are via Waterford.
    Staverton–Jersey = 1 hr   Staverton–Waterford = 1 hr 10 min
    Staverton–Guernsey = 1 hr 30 min   Staverton–Cork = 1 hr 30 min

5.

College Structure - College of Midshire

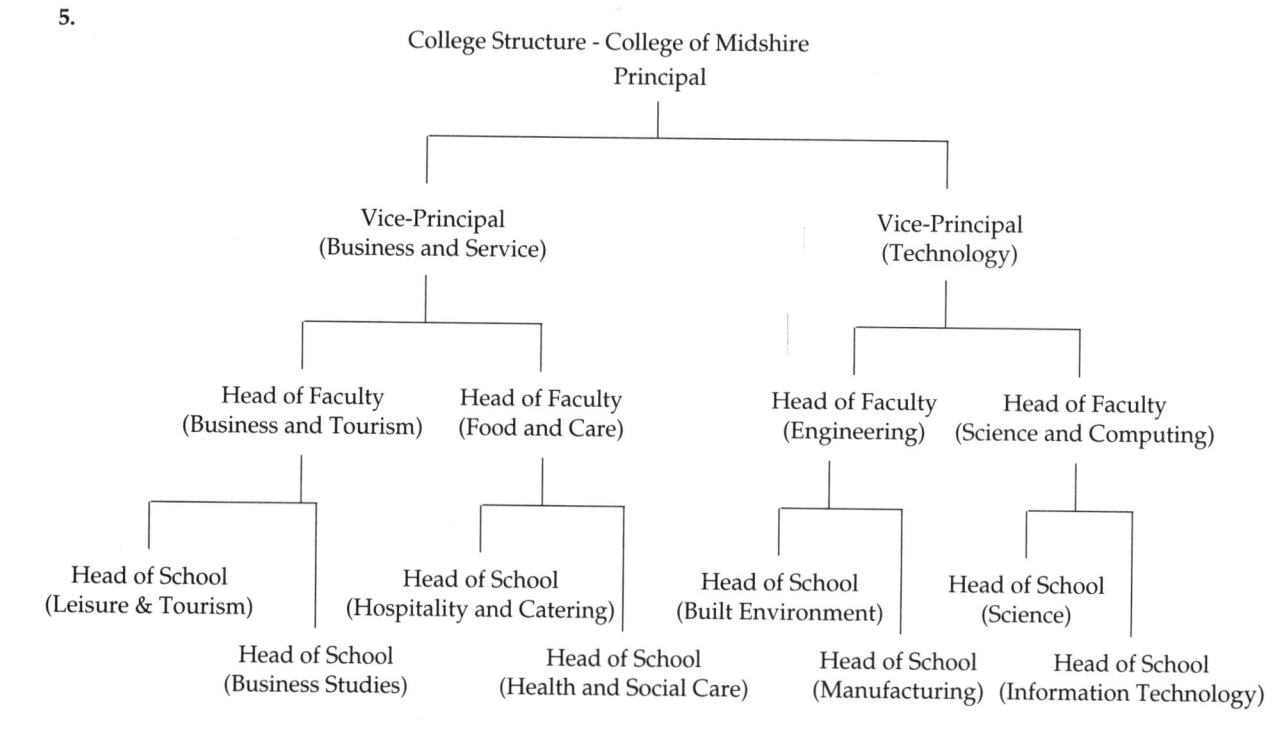

Principal
├── Vice-Principal (Business and Service)
│   ├── Head of Faculty (Business and Tourism)
│   │   ├── Head of School (Leisure & Tourism)
│   │   └── Head of School (Business Studies)
│   └── Head of Faculty (Food and Care)
│   ├── Head of School (Hospitality and Catering)
│   └── Head of School (Health and Social Care)
└── Vice-Principal (Technology)
    ├── Head of Faculty (Engineering)
    │   ├── Head of School (Built Environment)
    │   └── Head of School (Manufacturing)
    └── Head of Faculty (Science and Computing)
        ├── Head of School (Science)
        └── Head of School (Information Technology)

Chapter 15 Formulae and Equations p.249

1. $P = EI$
   (a) $P = 250 \times 8 = 2000$ watts
   (b) $P = 240 \times 0.5 = 120$ watts

2. (a) $l = 2450p$; l = lira, p = pounds
   (b) $y = 154p$; y = yen, p = pounds
   (c) $s = 1.45p$; s = dollars, p = pounds

3. (a) $B = 65d + 75$; B = bill, d = days
   (b) $B = 65 \times 6 + 75$
       $= 390 + 75$
       $= 465$
   (c) Weely bill = £325 + 75 = £400
       It would be cheaper to hire for one week.

4. (a) $B = 9w + 50$; B = bill, w = weeks
   (b) $B = 9 \times 6 + 50$
       $= 54 + 50$
       $= 104$
       Cost = £104

5. (a) $B = 7.50d + 30 + 9$; B = bill, d = days
       $= 7.50d + 39$
   (b) $B = 7.50 \times 3 + 39$
       $= 22.5 + 39$
       $= 61.5$
       Cost = £61.50

6. $C = \frac{5}{9}(F - 32)$; C = degrees centigrade, F = degrees Fahrenheit
   (a) $C = \frac{5}{9}(+32° - 32°) = 0$
   (b) $C = \frac{5}{9}(40 - 32) = \frac{5}{9} \times 8 = 4.44$ to 2 d.p.
   (c) $C = \frac{5}{9}(62 - 32) = \frac{5}{9} \times 30 = 16.67$ to 2 d.p.
   (d) $C = \frac{5}{9}(81 - 32) = \frac{5}{9} \times 49 = 27.22$ to 2 d.p.

7. (a) $C = 11.50l + 6.50$; C = cost, l = lessons
   (b) $C = 11.50 \times 5 + 6.50$
       $= 57.50 + 6.50$
       $= 64$
       Cost = £64

8. (a) $C = \pi d$; C = circumference, d = diameter
       $C = \pi \times 1$
       $C = \pi = 3.14$ Circumference = 3.14 m
   (b) $C = 38\pi$
       $= 119.38$ to 2 d.p.
       Circumference = 119.38 m

9. $E = 12$ V
   2 headlights at 60 W   = 120 W
   2 tail lights at 12 W   = $\underline{24}$ W
                        $P = \underline{\underline{144}}$ W

   $I = \dfrac{P}{E}$

   $I = \dfrac{144}{12} = 12$ amps

10. $P = \dfrac{(79.99 - 42)}{42} \times 100$

    $= \dfrac{37.99}{42} \times 100 = 90.45\%$

11. (a) $B = 19.54 + 0.0378u$; B = bill, u = units
    (b) 1st quarter $B = 19.54 + 0.0378 \times 937$
                      $= 19.54 + 35.42 = 54.96$
        2nd quarter $B = 19.54 + 0.0378 \times 1430$
                      $= 19.54 + 54.05 = 73.59$
        3rd quarter $B = 19.54 + 0.0378 \times 1199$
                      $= 19.54 + 45.32 = 64.86$
        4th quarter $B = 19.54 + 0.0378 \times 898$
                      $= 19.54 + 33.94 = 53.48$
    (c) Bill for the year = 54.96 + 73.59 + 64.86 + 53.48
                         = £246.89
    (d) VAT at 17.5% = £43.21
    (e) Total bill + VAT = £246.89 + £43.21 = £290.10

12. $I = \dfrac{PRT}{100}$

    $= \dfrac{2000 \times 6.3 \times 4}{100}$

    $= £504$

13. (a) $B = 10.10d + 1.47u$; B = bill, d = days, u = units
    (b) 1st quarter
        $B = 10.10 \times 102 + 1.47 \times 10\,715$
        $= 1030.20 + 15\,751.05$
        $= 16\,781.25$p
        $= £167.81$

        2nd quarter
        $B = 10.10 \times 94 + 1.47 \times 8700$
        $= 949.4 + 12\,789$
        $= 13\,738.4$p
        $= £137.38$

        3rd quarter
        $B = 10.10 \times 90 + 1.47 \times 3320$
        $= 909 + 4880.4$
        $= 5789.4$p
        $= £57.89$

        4th quarter
        $B = 10.10 \times 80 + 1.47 \times 3285$
        $= 808 + 4828.95$
        $= 5636.95$p
        $= £56.37$

Index

Algebra 239–42
Angle 180, 190
Area 179, 182–4, 191, 194, 199–202, 209, 219, 221
Average 154–8, 162–3

Bar charts 122, 136–143
BODMAS 24–8, 41, 47
Brackets 24–8

Calculator use 5, 15, 25, 27–9, 48, 62, 80, 82, 160
  memory 63–7
Cancelling 35, 41
Capacity 217
Centre 194
Charts 123–44
Circles 194–200
Circumference 194–9
Common factor 36, 41
Common denominator 37–9, 46
Conversion 88–97
  currency 97
  graphs 95–6
  tables 88–92
Cubes 209–11, 220
Cuboids 209–11, 220
Cylinder 209, 213–16, 217

Data 103–5, 156
  grouped 156
Decimals 48, 53, 55
  dividing 56–7
  multiplying 58
  subtracting 56
Decimal places 59, 81
Decimal point 53–9
Denominator 33, 35, 37–9, 40, 47
Diagonal 180
Diameter 194–9
Directed numbers 12, 14

Equations 239–41, 245–8
Equilateral triangle 187, 189

Factor 36, 41
Formulae 239–45
Fractions 32–6
  adding 37
  to decimals 48
  dividing 42–3
  equivalent 34–5
  improper 45
  multiplying 40
  proper 45
  subtracting 38
Frequency 116–19, 157, 158
  tables 116, 118, 119, 157, 158
  grouped tables 118–19
Graphs 95, 122, 133–44
  bar 122, 136–44
  line 133–4
Greater than 54

Imperial 88
Interviews 106–8
Isosceles 187, 189
Less than 54

Mean 154–8, 165
Measures of dispersion 165
Median 162
Metric 88
Mixed numbers 45–7
Mode 163

Negative numbers 13–19, 22
Net 218–21
Networks 226, 227, 229, 231–4
Numerator 33, 35, 40

Order of magnitude 7

Parallel 92–4, 180–4
  scales 92–4
Parallelogram 180–4
Percentages 72, 78–82
  decrease 81
  increase 81
Perimeters 179–87, 194, 209

Pictograms 122–8
Pie charts 122, 129–32
Place value 54
Probability 168–75
  tables 172

Quadrilaterals 180–2, 194, 200
Questionnaires 108–10

Radius 194–9, 215
Range 164–5
Ratios 72–4
Reciprocal 44–5, 65–7
Rectangle 180, 189, 190, 201
Relationship networks 231–4
Right angle 180, 188–91
  triangle 188, 191
Road maps 226–30
Rounding 2, 3, 59, 60, 62

Scales 72–8
Significant figures 59, 61, 62
Square 180
Surveys 102–4
  by interview 106
  by observation 105–6
Symbols 241–2, 245–8

Tables 113–19
Tally 116–18
  charts 116
Transport networks 229, 231
Trapezium 180–4
Tree diagrams 173
Triangles 186–93, 200, 201
  equilateral 187, 189
  isosceles 187, 189
  scalene 188
Triangular prisms 209–13, 220, 221

Units 183, 210
  Cubic 210
  Square 183

Volumes 179, 209–16